Cases

for Modern Systems Analysis and Design

To Timothy and Amanda

Cases for Modern Systems Analysis and Design

Annette Easton
George Easton
San Diego State University

The Benjamin/Cummings Publishing Company, Inc.

Menlo Park, California • Reading, Massachusetts • New York

Don Mills, Ontario • Harlow, U.K. • Amsterdam • Bonn • Paris

Milan • Madrid • Sydney • Singapore • Tokyo • San Juan, Puerto Rico

Executive Editor: Michael Payne
Senior Acquisitions Editor: Maureen Allaire
Assistant Editor: Susannah Davidson
Marketing Manager: Melissa Baumwald
Production Editor: Sue Purdy Pelosi
Manufacturing Supervisor: Casimira Kostecki
Cover Designer: Yvo Riezebos
Text Design and Composition: London Road Design
Copy Editor: Teresa Thomas
Proofreader: Cathy Linberg
Cover Art: La belle jardiniere, 1939, by Paul Klee, oil and tempera on burlap, Kunstmuseum Berne, Paul-Klee-Stiftung; © 1996 ARS, New York/VG Bild-Kunst, Bonn.

ISBN: 0-8053-2516-6

1 2 3 4 5 6 7 8 9 10 VG 00 99 98 97 96

The Benjamin/Cummings Publishing Company, Inc.
2725 Sand Hill Road
Menlo Park, California 94025

Preface

The goal of *Cases for Modern Systems Analysis and Design* is to provide information systems instructors and students with a wide range of realistic, current business situations appropriate for teaching and learning the methodologies, tools, and techniques of modern systems analysis and design. Each case evolved from an actual business situation, and, as such, the cases collectively address a broad range of systems analysis and design issues. This composition makes *Cases for Modern Systems Analysis and Design* adaptable to the range of teaching styles and course objectives found today in undergraduate and graduate systems analysis and design courses across the country.

The sources of these cases are real organizations and real people. To retain the realism, we have tried to present the material for each case as it was originally provided by the client. A challenge of systems development is coping effectively with the unrealistic expectations, imperfect details, and extraneous information provided by clients. This realism is instructive in helping students appreciate the modern-day issues of systems development.

We have purposely avoided setting specific milestones for the cases, since we do not want to constrain instructors from setting their own learning objectives when using the cases. We encourage instructors to apply the methodologies, tools, and techniques of systems analysis and design taught in their courses to the cases in this book. Instructors may find that the explicit learning objectives highlighted by the questions at the end of each case can be complemented by topic-specific objectives by making plausible assumptions about the cases. For example, a cost–benefit analysis may be developed for a case by using current estimates of development costs (e.g., cost of hardware and/or cost of consultant) and credible estimates of the benefits (e.g., labor savings). This format also allows instructors to modify the case requirements based on the time constraints imposed by their course schedule. Classes that have less time for projects can reduce or modify the requirements, while classes that spend a greater time on projects can more fully develop prototypes and project plans.

LEARNING OBJECTIVES OF THE CASES

A summary of the learning objectives and suggestions for applying each case to the systems development process follows.

1. Hylton's

Hylton's is a case considered appropriate for the individual student or a small group of students just beginning the IS curriculum. The primary learning objectives of the case are problem definition, project scope, data flow diagramming, and initial feasibility assessment. The Hylton's case may also be appropriate for students to use as a project for prototyping. Beginning IS students can use this case to demonstrate how payroll processing activities, such as those at Hylton's, can be made more efficient, even with a simple tool such as a spreadsheet.

2. Picnics Plus

The Picnics Plus case is appropriate for a team of IS students who have had exposure to the concepts of systems analysis. The learning objectives associated with this case include problem definition, project scope, and data flow diagramming. The Picnics Plus case may also be appropriate for use as a prototyping project.

3. KPUB

The KPUB case can be used for both analysis and design-related activities. For example, instructors can ask their students to develop a series of questions that could be used in interviewing KPUB staff in order to further the analysis process. Additionally, students have been given sufficient information to develop level-0 DFDs of KPUB's existing system. Students may also be able to develop the entities of an E-R diagram based on the information given. This case may also be used to help develop the concepts of change management and business process reengineering.

4. Media Technology Services (MTS)

The MTS case can be used for both analysis and design-related activities. It is most appropriate for teams of students who have had exposure to the concepts of systems analysis. The case describes a messy situation in which a system has been developed without following any structured techniques. Students have been given enough information to develop DFDs of MTS's existing system. Students should also be able to create an E-R diagram and a set of normalized relations. Based on the requirements of the case, student teams should be able to develop a prototype of a new system. The case requires students to be concerned with issues of data conversion from a prior system, as well as looking at issues of future goals.

5. Meeting Makers

The Meeting Makers case can be used by either small teams or individual students. It covers the range of analysis and design activities. The company does not have any in-house technology expertise. They are using (and misusing) existing programs. Students should be able to create DFDs, develop normalized relations, build a system prototype, and reengineer procedures. Additionally, students are asked to address issues relating to user training, Internet usage, and scanning technology.

6. Homeowners of America (HOA)

The Homeowners of America case is best suited as a complete project for an individual student. While the focus is on the analysis activities, it does allow for prototyping on a smaller scale. The case provides the opportunity for creating data flow diagrams, process reengineering, system design, and prototypes. It can also be used to develop a midrange hardware/software technology upgrade plan.

7. Hazardous Materials Management System (HMMS)

HMMS can be used as either a systems analysis case or a systems design case. It is most appropriate for a team of students who have had exposure to systems analysis. Students should be able to develop data flow diagrams, entity-relationship diagrams and normalized relations. The case offers students a chance to focus on many of the "people" issues of analysis and design such as deciphering user requests, centralization/decentralization issues, implementation change management, and policy/procedure design. These tasks could be accomplished even by students not thoroughly trained in the more technical areas of development.

8. National Booksellers, Inc. (NBI)

The NBI case can be used by either teams or individuals. This case covers the full range of analysis and design activities, including data flow diagrams, entity-relationship diagrams, cost–benefit analysis, and prototyping. It provides an opportunity to deal with sites in different parts of the country. Students can also explore issues relating to developing systems for a subsidiary of a larger organization, while allowing for the possibility of eventual merging into the parent company.

9. TechPrint, Inc.

TechPrint, Inc. provides individual students or a small team of students with the opportunity to develop the business case for a World Wide Web presence. The issues of this case are the same issues confronting organizations of all sizes today. Primary learning objectives of this case are the development of a baseline project plan and statement of work.

10. Red Rock City

The Red Rock City case is best suited as a complete project for either an individual student or a student team. The focus of the case is on systems development activities including database file design, screen layouts, report layouts, and implementation plans. This case is appropriate for prototyping and incorporates integrating data from external entities.

ACKNOWLEDGMENTS

Many people have contributed to the completion of this casebook. First, we would like to thank Jeff Hoffer, Joe Valacich, and Joey George for asking us to be involved in this project. Their direction and feedback have helped in refining the cases included in this text.

We have both benefited from the contributions of a variety of students who dedicated long hours to solving early versions of the projects and cases. Many thanks to the students in our systems analysis, systems design, database management, and practicum courses.

The reviewers provided insightful comments and feedback that helped to refine the clarity and writing of the cases. Our appreciation is extended to Robert B. Jackson, Brigham Young University; Mary B. Prescott, University of South Florida; Eugene F. Stafford, Iona College; and Connie E. Wells, Nicholls State University.

Finally, we would like to thank the editorial and production staff at Benjamin/Cummings. Special thanks go to Maureen Allaire and Susannah Davidson for providing encouragement and guidance and for helping us wrap up this project.

Table of Contents

Hylton's

Hylton's, a popular beach-area restaurant, was started five years ago by Nann Hylton as a small catering business. After five years as a catering operation, Nann had the opportunity to expand to a full-scale restaurant. Her business has been successful, in part, because she has adhered to the central business principle she learned long ago: offer the customer a good meal at a fair price. However, Nann realizes that she must alter some of her operations as she moves to a full-service restaurant business.

One operational activity that Nann is considering changing is the process the firm uses for payroll. With approximately 25 new employees being added to Hylton's existing payroll of 10, the change in the workload of the people responsible for payroll is significant.

HYLTON'S PAYROLL SYSTEM

The present payroll system used by Hylton's employs both automated and manual processes. The automated processes, however, are either data sources for the manual payroll activities done at Hylton's or serve as destinations for Hylton's payroll processing activities.

An automated source of data for the payroll process at Hylton's is their NCR 2160 cash register. This machine was primarily intended for use as a cash register, but it also maintains a number of data files that are potentially useful for other business functions, including one for employee timekeeping.

At Hylton's, employees initiate the daily timekeeping process when they begin their shift and clock in via the 2160; they also use the machine to clock out at the end of their shift. The employee timekeeping data file generated by the NCR 2160 can store up to one week of time-keeping data. However, Nann has not explored using the 2160 timekeeping data more fully because she pays her employees every two weeks.

At the close of each business day, the operations manager, Kristin, reviews a daily time-keeping tape from the 2160. Kristin checks the tape for errors (e.g., an employee forgets to clock in or clock out), and makes the necessary adjustments to the employee timekeeping records on the 2160.

At the end of the week, when all data are adjusted and correct, the data are printed to a weekly labor summary tape (Figure 1-1). Next, Kristin reviews the labor summary tape and rounds the weekly employee hours worked to the nearest quarter hour. Regular and over-time hours are also calculated. Kristin then transfers this data manually to a weekly time sheet (Figure 1-2) provided by PayChex, the company that processes Hylton's payroll. Kristin had been spending approximately 5 hours every week reviewing, adjusting, and preparing the weekly time sheet when the firm had only 10 employees. She is spending at least 15 hours each week on this process now that the firm has so many new employees.

Because Hylton's employees are paid every two weeks, these activities must be performed each week. At the end of the second week Kristin manually combines both weeks' data on a master time sheet (a photocopy of the weekly time sheet). Kristin spends an additional 5 hours every other week putting together the master time sheet. The master time sheet is then given to Nann, who makes the remaining payroll adjustments for the pay period before sending the time sheet to PayChex.

The adjustments Nann makes to the master time sheet are usually for bonuses, health insurance, uniforms, and advances. Nann spends about 2 hours each pay period making these adjustments to the master time sheet before she faxes it to PayChex. PayChex then completes its processing and sends the paychecks and various reports including the payroll journal and an employee earnings record, along with an updated time sheet, back to Hylton's. Nann finishes the payroll process for the period by verifying the calculations, signing the paychecks, and distributing them to the employees.

PayChex is a national payroll service provider that calculates and prints the actual employee paychecks, creates and updates the company's payroll journal, and provides any government reports needed. The time sheet is the form and format required by PayChex. PayChex charges Hylton's approximately $130 per month ($65 per pay period) for processing the payroll of the 35-employee firm. Since Nann is quite satisfied with the service provided by PayChex and because she does not want to have to worry about the periodic governmental reports that need to be filed, she does not plan to change the PayChex portion of Hylton's payroll processing. However, she does want to reduce the time both she and Kristin spend on payroll. Both Kristin (at $15/hour) and Nann (at $50/hour) are too valuable to spend as much time as they do on the payroll process.

This has prompted Nann into investigating the usefulness of the timekeeping database stored in the NCR 2160. She was once shown how to export the 2160 databases to her PC spreadsheet by the sales representative, and she was given all of the documentation on how the databases store the data (Figures 1-3 and 1-4). This information wasn't interesting to Nann until now because she was too busy growing the business. Furthermore, Nann is becoming much more comfortable using her PC (an IBM-compatible computer with an 80486 microprocessor running at 66MHz, 8MB of RAM, and a 300MB hard drive) and laser printer.

Labor Summary			10/8/95
Purdy, John	1	Hours	Dollar Cost
001-00-0003	JC	25:50	
Reg. Wages	509		
P.T.D. Tips			
Total Labor Cost			
Cooke, Isabella	1	Hours	Dollar Cost
002-00-0004	JC	25:50	
Reg. Wages	509		
P.T.D. Tips			
Total Labor Cost			
Iavicoli, Rita	1	Hours	Dollar Cost
003-00-0005	JC	25:50	
Reg. Wages	509		
P.T.D. Tips			
Total Labor Cost			
Peterson, Michael	1	Hours	Dollar Cost
004-00-0006	JC	25:50	
Reg. Wages	509		
P.T.D. Tips			
Total Labor Cost			
Crawford, Robert	1	Hours	Dollar Cost
005-00-0007	JC	25:50	
Reg. Wages	509		
P.T.D. Tips			
Total Labor Cost			
Salmon, Vicky	1	Hours	Dollar Cost
006-00-0008	JC	25:50	
Reg. Wages	509		
P.T.D. Tips			
Total Labor Cost			
Fisher, Nathan	1	Hours	Dollar Cost
007-00-0009	JC	25:50	
Reg. Wages	509		
P.T.D. Tips			
Total Labor Cost			
Garcia, Daniel	1	Hours	Dollar Cost
008-00-0010	JC	25:50	
Reg. Wages	509		
P.T.D. Tips			
Total Labor Cost			
Gould, Janet	1	Hours	Dollar Cost
009-00-0011	JC	25:50	
Reg. Wages	509		
P.T.D. Tips			
Total Labor Cost			

Figure 1-1   Weekly labor summary tape

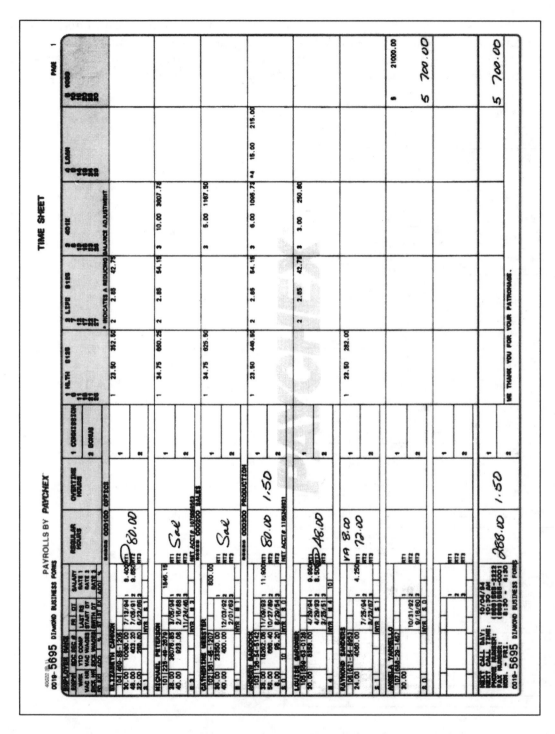

Figure 1-2 Weekly time sheet *(Courtesy of PayChex, Inc.)*

Field	Width/Characters
Employee Number	10
Employee Name	12
Time Card Number	6
Job Code 1	4
- Total Time - (HH)	4
- Total Time - (MM)	2
- Rate	9
- Over Time Factor	5
- Allowance 1 Amount	7
- Allowance 2 Amount	7
Job Code 2	4
- Total Time - (HH)	4
- Total Time - (MM)	2
- Rate	9
- Over Time Factor	5
- Allowance 1 Amount	7
- Allowance 2 Amount	7
Job Code 3	4
- Total Time - (HH)	4
- Total Time - (MM)	2
- Rate	9
- Over Time Factor	5
- Allowance 1 Amount	7
- Allowance 2 Amount	7
Period to Date Tips Amount	11

Figure 1-3   Record layout, employee timekeeping file

469851205, "Cannon, E," 000000,0030,0064,01,00005.25,1.500,+000.00,+000.00,0010,0064,01,
00005.500,1.500,+000.00,+000.00,0030,0064,01,00005.25,1.500,+000.00,+000.00,+0000000.00

Figure 1-4   Example of the NCR 2160 employee timekeeping record for one employee

Exercises

1. Define the problem(s) of the payroll system at Hylton's.

2. Write a statement of scope and objectives that Nann Hylton can use to evaluate whether your firm can help her solve these problems.

3. What is your initial assessment of the feasibility of solving these problems based on these objectives?

4. Draw a data flow diagram of the current payroll processing system at Hylton's.

Picnics Plus

Picnics Plus is the name given to the catering operation owned by Claudia Hennessey. Ms. Hennessey started Picnics Plus five years ago after managing a local restaurant that occasionally catered local businesses' summertime picnics. Picnics Plus is now a year-round operation, but summer picnics represent, on average, approximately 70 percent of the firm's annual revenue. During a typical summer weekend, Picnics Plus caters 1 to 10 picnics that range in size from 10 to more than 1,000 people. Picnics Plus is well-known for its menu and for its reasonable prices. Recently, Picnics Plus began generating new business prospects by providing food service at some of the more popular local summertime events and attractions, such as the symphony and the city's harbor cruise.

Ms. Hennessey is aware that Picnics Plus' existing process of setting up picnics may inhibit the catering business' growth. She is anxious to learn about alternative ways to conduct her business. She is especially interested in learning better ways to improve the response time for customer cost estimates, better ways to generate picnic pick lists, and better ways to get all of the cost information together after an event. She is also very interested in ensuring that there is no reduction in service, and she is adamant that any system adopted must be economically feasible and easy to use.

PICNICS PLUS' CURRENT OPERATION

In order to ensure proper ordering and preparation of food and supplies for each client, Picnics Plus has established an operational routine that, while hectic during peak periods, works successfully if procedures are followed. The process essentially involves the following people: Ms. Hennessey, the operations manager, the catering manager, and the kitchen manager.

Ms. Hennessey

Potential customers usually phone Picnics Plus with a request for a price quote for a picnic. The number of people attending the picnic, the menu information, and the customer information are written on a prospectus referred to as a "party sheet" (Figure 2-1) along with any special notes about the picnic. Customer information may already be on file at Picnics Plus if the client has used Picnics Plus previously. If such a file exists, it is pulled by the person taking the order and routed with the party sheet through the process.

The party sheet, as well as the customer file if available, are given either to Ms. Hennessey or to the catering manager to determine the picnic cost estimate and to review the event information for scheduling conflicts and other potential problems. After these issues have been addressed, Ms. Hennessey or the catering manager generates a quotation based on the expected attendance, the menu chosen, and the relationship with the client (for instance, repeat clients may receive a discount).

The menu used by Picnics Plus simplifies the initial effort necessary to provide a price quote for the customer and also simplifies the process for calculating the amount of food and supplies, personnel, and equipment needed for a picnic.

Customers can choose food from one of two menus, the Classic or the Classic Plus. The primary difference between the menus is the entrée: the Classic Plus menu includes steak or chicken, while the Classic menu includes a hamburger or a hot dog. Side dishes, drinks, and desserts are the same for both menus.

	Picnics Plus		
	Party Sheet		
Client:	General Dynamics	**Event Date:**	9 April 97
Address:	Information Systems Division	**Event Hours:**	11 a.m.–4 p.m.
	100 Pacific Highway	**Begin Food Service:**	Noon
	San Diego, CA 92110	**Location:**	Mission Bay
Phone:	619-555-1234		
Fax:	619-555-2234		
Contact:	Hugh Grant	**Sales Person:**	Melissa McDill
No. of Guests:	400 **Classic:** 400	**Classic Plus:**	
Notes:			

Figure 2-1   Picnics Plus party sheet

Ms. Hennessey's and the catering manager's hectic schedules sometimes cause delays in completing potential customer requests for quotations. This inevitably causes some potential customers to go elsewhere and tarnishes the image of Picnics Plus. This is a situation that Ms. Hennessey feels must be corrected soon, especially if she wants the business to continue to grow.

When the price quotation is complete, it is phoned to the prospective client. Once a client agrees that Picnics Plus will cater the event, Ms. Hennessey or the catering manager (usually whoever quoted the job) creates a contract from the approved party sheet and sends it to the client for a signature. The contract is a standard boilerplate form with fields filled with information from the party sheet. Copies of the contract and party sheet are put in the customer's file.

Catering Secretary

When the signed contract is returned, the catering secretary files the contract in the customer file and copies and distributes the party sheet to the catering manager (one copy), the operations manager (one copy), the kitchen (three copies), and the accounting department (one copy). Each department is responsible for completing its part of the picnic order. Three days prior to an event, the catering secretary contacts the customer to confirm the number of guests. If necessary, she will update the party sheet if any significant details of the event have changed; she will also notify other Picnics Plus staff of the changes.

Catering Manager

The catering manager uses the party sheet to order the food and supplies for all picnics. The menu system employed by Picnics Plus allows the catering manager to calculate the quantity of food and supplies to order for a given picnic based on attendance and on a unique multiplier for each menu item. For example, a picnic of 400 people will use, on average, 600 plastic plates (the plate multiplier is 1.5). The individual multipliers were determined some time ago by Ms. Hennessey and considered in the pricing of picnics. There are similar multipliers for the number of tents, tables, trucks, and other equipment that is reserved by the operations manager.

The catering manager creates a food and supplies sheet (Figure 2-2) for each picnic. The catering manager then uses the individual food and supplies sheets to help in ordering all of the supplies for upcoming picnics. After the event, the catering manager forwards the party sheet and the food and supplies sheet to the accounting manager.

Picnics Plus
Food and Supplies Sheet

Client:	General Dynamics	Event Date:	9 April 97
Address:	Information Systems Division	Event Hours:	11 a.m.–4 p.m.
	100 Pacific Highway	Begin Food Service:	Noon
	San Diego, CA 92110	Location:	Mission Bay
Phone:	619-555-1234		
Fax:	619-555-2234		
Contact:	Hugh Grant	Sales Person:	Melissa McDill

No. of Guests: _____400_____   **Classic:** _____400_____   **Classic Plus:** _____

Item Description	Quantity	Item Description	Quantity
Hamburgers	400	Hamburger Buns	400
Hot Dogs	400	Hot Dog Buns	400
Swiss Cheese	200 pcs	American Cheese	200 pcs
Lettuce Cups	20 heads	Sliced Tomatoes	100 whole
Sliced Onions	40 whole	Chopped Onions	20 lbs
Dill Pickles	20 lbs	Sweet Pickles	20 lbs
Celery Sticks	16 lbs	Carrot Sticks	16 lbs
Tossed Green Salad	20 lbs	Ice Cream Cups	400
Lemonade	400 people		
10.5" Plastic Plates	600	7.5" Plastic Plates	600
Ind Plastic Forks	800	Ind Plastic Knives	400
Ind Plastic Spoons	400	Paper Napkins	2000
9 oz Cold Cups	1200	9 oz Serco Plastic	400
Solo's	1200	Silver Forks	12
Silver Spoons	12	Silver Knives	12
Grill-Pam-Gas	2	Barbecues	2
Charcoal	80 lbs	Lg Chafers w/inserts	4
Sm Chafer w/inserts	4	Sm Throwaways	8
Lg Throwaways	8	Sterno	48 cans
Lg Slot Serving Spoon	2	Lg Solid Serving Spoon	2
Sm Solid Serving Spoon	4	Lg Tongs	4
Sm Tongs	4	Lg Ladle 6 oz	4
Sm Ladle 2 oz	4	Long Spatula	2
Aluminum Foil	2	Carving Setup	2
3 Lite Bar	2	Barbecue Brush	2
Lemonade Cooler	2	Iced Tea Cooler	2
Trash Bins	6	Trash Bags	20
Sugar	400	Sweetener	400
Stirrers	400	Instant Tea	400
Lemon Packages	400	Ketchup Packages	600
1000 Island Dressing	400	Mustard Packages	400
Relish Packages	400	Mayo Packages	400
Disposable Salt	400	Disposable Pepper	400
Ice Bin Brown	4	Linen	32
Ice Walnut 50 lbs	2	Iced Tea	50 gal
Dry Ice Sliced	2	Tool Box	2
Knife for Cake	2	8 Ft Tables	8
6 Ft Tables	4		

Figure 2-2   Food and supplies sheet

Operations Manager

The operations manager schedules the event on the event calendar. The operations manager also prepares an equipment list of specific items that must be rented for the picnic: for example, large tents, chairs, tables, and truck(s) to deliver the picnic supplies. As much as possible, the equipment information for a picnic is aggregated with the equipment information needed for other picnics during that time period. The operations manager tries to use the aggregated equipment information to obtain optimal pricing for the equipment from a local party equipment rental outlet. The equipment list is forwarded to the accounting department after the event.

The operations manager is also responsible for scheduling staff for the event or picnic. The number of cooking, serving, and other staff required are estimated based on the number of people attending the picnic. As much as possible, specific Picnics Plus staff are assigned to the picnic at this point. If the firm continues to grow, staff scheduling may become more of a problem than it is now. Currently, Picnics Plus has what seems to be an ample supply of labor since many college students, home for the summer, find the work, hours, and pay of Picnics Plus desirable.

Kitchen/Beverage/Pantry Managers

The kitchen, beverage, and pantry managers are responsible for the preparation of the food and beverages ordered for each picnic. The kitchen manager oversees the staff of people who prepare and load the hot food for each picnic on the day of the picnic. The beverage manager is responsible for the people who prepare and load the beverages. The pantry manager is responsible for the preparation and loading of all the cold food.

Accounting Manager

The accounting manager receives the completed party sheet, equipment list, and contract to prepare the customer's invoice. The accounting manager is also responsible for gathering the data and billing for any special equipment requested for the event. These charges are itemized on the invoice, which is then sent to the client.

ADDITIONAL INFORMATION

Computer Equipment

Picnics Plus has access to Ms. Hennessey's microcomputer, an IBM-compatible computer equipped with an 80486 DX2 microprocessor running at 40MHz with 8MB of RAM and a 300MB hard drive that is approximately half full. Ms. Hennessey also has a laser printer and a word processing application.

Exercises

1. How would you define the problem at Picnics Plus?

2. Create a statement of scope and objectives that addresses the issues Ms. Hennessey would like to see corrected at Picnics Plus.

3. Draw a data flow diagram of the current operation at Picnics Plus.

4. What questions must you ask of Ms. Hennessey in order to complete an initial feasibility assessment?

KPUB

K PUB is a public television and radio station affiliated with the University of California. Until a few years ago, all of KPUB's departments were located on campus. However, as the university's need for space grew, KPUB was forced to relocate its telecommunications department, engineering department, and business office in separate buildings a short distance from the campus. These functional units, plus the production department, which is still on campus, comprise the organizational structure at KPUB.

KPUB, like many public broadcasting stations, has sought supplementary sources of funding for its operations. Many stations are now using excess capacity to provide telecommunications services for private organizations and government agencies. These services at KPUB include teleconferencing, video production, and audio production.

The billing process for telecommunications services worked adequately when KPUB's clients were all internal. This was true, in part, because KPUB had no specific time limit to produce invoices, nor was the station required to adhere to any formal audit procedures. Often, it would take up to three months for a complete invoice to reach an internal customer.

The process of tracking the expenses of a telecommunications conference or a video production is generally considered responsible for the delay in the billing process at the station. The billing process is further complicated by the geographical location of project activities at KPUB and by a hectic environment that places more emphasis on impending and current projects than on completed projects.

Now that KPUB's client list includes outside organizations, KPUB must be more prompt and accurate in its billings. Furthermore, since KPUB has also started providing services for government agencies, the station must now adhere to more rigid audit requirements.

EXISTING SYSTEM

KPUB's current telecommunications billing process begins when a request is made to the telecommunications department for a telecommunications conference or a video production. For viable projects, the telecommunications department initiates a bid based on an estimate of the total costs of the project. Bids are determined using a multi-page budget request form that itemizes all of KPUB's expense categories for its internal personnel (Figure 3-1) and facilities usage (Figure 3-2) charges; and for the external supplies and travel (Figure 3-3), operating, subcontracted, and contracted services expenses (Figure 3-4) related to the project.

Production Budget Request Summary—Internal
PERSONNEL

Code	Description	Rate	Budget Units (Hours)	Total Dollars
6200	PERSONNEL BUDGET			
6205	Executive Producer	$50		
6360	Senior Producer	$38		
6210	Producer	$28		
6150	Associate Producer	$23		
6220	Research Assistant	$16		
6230	Director	$29		
6380	Field Camera	$24		
6170	Production Technician	$14		
6180	Video Production Engineer	$24		
6190	ITFS Production Technician	$15		
6250	Audio Production Engineer	$23		
6270	Production Assistant	$8		
6390	Studio Supervisor	$40		
6370	Assistant Studio Supervisor	$20		
6140	Engineer	$30		
6410	Senior Engineer	$32		
6160	Telephone Operator	$18		
6350	Production Coordinator	$43		
6420	Telecommunications Coordinator	$26		
7331	Art/Photo Services	$30		
7342	Promotion Services	$31		
7345	Engineering Consulting Services	$41		
7344	Telecommunications Consulting Services	$23		

Total Personnel _____

Figure 3-1   Budget request form: Personnel

Production Budget Request Summary—Internal
FACILITIES

Code	Description	Rate	Budget Units (Hours)	Total Dollars
7300	FACILITIES BUDGET			
7301	TV Classroom (includes 1 operator)	$77		
7304	Audio Studio Recording (no personnel)	$2		
7311	VTR Playback/Screen	$33		
7312	VCR Playback/Screen	$6		
7313	VTR Dubbing—1"	$47		
7314	VCR Dubbing—3/4"	$47		
7336	VCR Dubbing—1/2"	$47		
7348	Window Dubbing—1"	$47		
7349	Window Dubbing—3/4"	$47		
7338	Spot Dubbing (up to 60 seconds)	$1/ea.		
7315	Video Edit—Computer (includes 1 Video Engr)	$147		
7317	Video Edit—Off-line (no personnel)	$45		
7325	Limited Video Production (DVE Programming)	$160		
7326	Full Video Production	$277		
7327	Studio Prep/Strike	$79		
7328	Field Video Production—1" (no personnel)	$47		
7335	Field Video Production—3/4" (no personnel)	$45		
7337	Sound Stage Rental (TV Studio)	$15		
7329	CG/TelePrompTer Program (includes 1 operator/programmer)	$66		
7330	Film Services (no personnel)	$12		
7343	Extra Equipment Services (see Prod. Coord.)			
7347	Graphics Production	$106		
7350	Downlinking	$94		
7351	Satellite Xmission/ITFS Distribution	$47		

Total Facilities _____
Total Personnel _____
Total Internal Expenses _____

Figure 3-2   Budget request form: Facilities

Production Budget Request Summary—External
SUPPLIES

Code	Description					Total Dollars
6700	SUPPLIES BUDGET					
6755	Staging/Props					
6756	VTR Stock:					
	Quantity:	Type:	Length:	@ $		
	Quantity:	Type:	Length:	@ $		
6757	VCR Stock:					
	Quantity:	Type:	Length:	@ $		
	Quantity:	Type:	Length:	@ $		
	Quantity:	Type:	Length:	@ $		
6758	Film Supplies (Miscellaneous)					
6759	Film Stock					
6760	Audio Supplies (audiotape, 16mm mag.)					
6797	Other Supplies					

Total Supplies _____

Production Budget Request Summary—External
TRAVEL

Code	Description					Total Dollars
6800	TRAVEL BUDGET					
6801	Local Mileage:					
		miles @		per mile + parking $		
6802	California Travel			days @		per day	
6803	U.S. Travel			days @		per day	
6804	Foreign Travel			days @		per day	
6820	Consultant Travel			days @		per day	
6821	Participant Travel			days @		per day	
6897	Other Travel					

Total Travel _____

Figure 3-3   Budget request forms: Supplies and travel

Production Budget Request Summary—External
OPERATING, SUBCONTRACTED, AND CONTRACTED EXPENSES

Code	Description	Total Dollars
7000	OPERATING EXPENSE BUDGET	
7010	Postage (Express Mail)	
7011	Shipping	
7019	Facsimile Transmission	
7023	Film Lab Processing	
7024	Duplicating	
7025	Printing (Promotional)	
7030	Equipment Rental	
7031	Vehicle/Vessel Rental	
7034	Facilities Rental	
7041	Public Relations	
7044	Advertising	
7046	Food Services	
7054	Rights	
7404	Negative Cutting	
7405	Prints/Opticals	
7406	Work Prints	
7407	Tape Duplicating	

Total Operating Expenses _____

Code	Description	Total Dollars
8500	SUBCONTRACTS	

Total Subcontracted Expenses _____

Code	Description	Total Dollars
8600	CONTRACTED SERVICES	
8601	Consultant Fees	
8602	Consultant Expenses	
8611	Professional Services	
8612	Technical Services	
8616	Talent	

Total Contracted Expenses _____

Recap External Costs

Code	Description	Total Dollars
	TOTAL DIRECT EXPENSES	
8900	INDIRECT COST-FDN @ 6%	
	TOTAL SUPPLIES	
	TOTAL TRAVEL	
	TOTAL OPERATING EXPENSES	
	TOTAL SUBCONTRACTED EXPENSES	
	TOTAL CONTRACTED EXPENSES	

Total External Expenses _____

Figure 3-4   Budget request form: Operating, subcontracted, and contracted expenses

The budget request form and bid process begin in the telecommunications department, where the project request originates. The appropriate budget request forms are then sent to the production department and then the engineering department with a copy of the project proposal to complete the bid. The telecommunications department estimates such things as site costs, telephone and supply expenses, producer expenses, and interconnect charges. The production department estimates director expenses and studio crew expenses. The engineering department estimates charges such as operating technicians, special equipment, and maintenance.

After the costs are estimated, the project proposal and budget request forms are sent to the business office, where a contract is created. The contract is then sent to the client for signature. The signed contract and a 50 percent deposit put the project on KPUB's production schedule.

When the signed contract is returned to the business office, the pertinent project information (copy of the proposal, copy of the production schedule, and copies of the departmental estimates) are distributed to the telecommunications, production, and engineering departments.

During preproduction, production, and post-production of the project, the staff of each department fills out cost information sheets (Figure 3-5) when they work on, use supplies for, or use a facility for a project. The cost information sheets are intended to be filled out as charges are incurred. At the conclusion of a project, the department managers are asked to forward the cost information sheets to the business office, where they are organized and used to generate an invoice.

For some time, the departments have tried to cross-check the cost information sheets for accuracy before sending them to the business office because much of the incoming charge-related paperwork has been suspect. For example, employees recording facilities charges do not have time to enter them in the logs until days or weeks after the job. When they finally do record the charges, the accuracy of the entry is limited to the employee's memory of what she/he worked on weeks earlier.

Cross-checking project charges helps improve the accuracy of KPUB's billing process but significantly slows the creation of client invoices. Also slowing the billing process are the charges incurred at remote sites, which are generally not submitted until after the site has been broken down, and the vendor charges that are chargeable to the client but not always available in a timely manner. (This is partly due to employees forgetting to inform vendors of the account code associated with a particular purchase.)

Cost Information Sheet

Project: _____ Day: _____ Date: _____ Client: _____

Program: _____ P.O.#: _____ Producer: _____

FACILITIES	HRS	RATE	TOTAL	INT	PERSONNEL	HRS	RATE	TOTAL	INT
Studio Prep/Strike					Exec. Producer				
Full Video Production					Producer				
Limited Production					Director				
Modified Production					Moderator				
C.G. Programming					Research Ass't/Writer				
TelePrompTer Programmer					C.G. Programmer				
Studio Audio Record					Associate Producer				
Studio Only					Assistant Director				
Set Construction					Additional Audio				
Video Edit/Computer					Additional Engineer				
Video Edit/Manual					Field Cam/Editor				
Video Edit/Off-line					Field Audio				
VTR Dubbing—1″					Prod. Assistant/Grip				
VCR Dubbing—3/4″					Set Designer				
VCR Dubbing—1/2″					Telecom. Coordinator				
VTR Screening					Phone Operator				
VCR Screening					Prompter Prog.				
Art/Photo Services					Editor				
TV Classroom					Senior Engineer				
El Nido Film/Sound					Assistant Studio Supr.				
Receive Off Satellite									
Tape Off Satellite									
TOTAL FACILITIES					TOTAL PERSONNEL				

EXTERNALS	QTY	RATE	TOTAL	INT	NETWORK/SATELLITE	TIME	RATE	TOTAL	INT
Video Tape—1″, 2 hrs					Transponder				
Video Tape—1″, 1 hr					Uplink				
Video Tape—1″, 20 min					Microwave				
UCA-60					Site				
UCA-30					Site				
UCA-20					Site				
Video Tape—1/2″					Site				
Audio Tape Supplies					Site				
Set Construction Materials					Site				
Telephones					Site				
Mileage					Site				
Mileage					Site				
Equipment Rental					Site				
Catering					Per Site Charge				
Miscellaneous									
TOTAL EXTERNAL					TOTAL NETWORK				

Figure 3-5   Cost information sheet

KPUB management is concerned about the delays encountered in producing invoices and the resulting delay in KPUB receiving payment for services. Additionally, they would like to be able to better monitor charges on projects so their estimates can be more realistic. Currently their estimates are low (approximately 20 percent below the actual cost) 25 percent of the time. Since contracts are based on the estimated amount and allow only for a maximum 10 percent overcharge, KPUB has the potential to lose revenue. As the quantity of outside work increases, this is even more of a concern.

KPUB is very interested in using its computer equipment in the bidding and billing processes but has no immediate plans for adding permanent staff to maintain its computer system. The station has one 386 microcomputer with 4MB of RAM running Windows and Microsoft Word in each of its production, engineering, and telecommunications departments. All are equipped with 120MB hard drives. Additionally, KPUB's billing office has two other microcomputers: a 386 configured similarly to the other KPUB machines (except this machine also has a spreadsheet package) and a Macintosh 8100 PowerPC. None of the machines at KPUB are networked.

The permanent staff in KPUB's three departments are comfortable using the word processing capabilities of the equipment, but generally these employees only use the equipment for writing memos. The billing office permanent staff is also comfortable using the spreadsheet package loaded on its machine.

Exercises

1. From the information given, define the problem(s) at KPUB.

2. Create a statement of scope and objectives for the KPUB problem.

3. Develop a set of questions to be used in an interview with KPUB management that are important in the analysis of KPUB's bidding and billing processes.

4. Draw a DFD of the existing billing system.

5. Develop a set of procedural recommendations that KPUB can implement to improve the accuracy and timeliness of the current billing process.

6. Develop a proposal on how KPUB can better use information technology to improve the billing process.

Media Technology Services (MTS)

Media Technology Services (MTS) is a service department at Desert View Community College. Desert View is one of the largest community colleges in the Southwest, with several different campuses spread over a 60-mile range. Last year the college served more than 25,000 students. One of Desert View's goals is to be a leader in the use of instructional media to support education. The MTS department was created to help the college achieve that goal. MTS provides support to faculty and staff in the acquisition and distribution of instructional media, in the distribution and maintenance of audiovisual equipment, and in the design of instructional material. MTS served more than 1,500 faculty and staff last year. They currently offer more than 7,500 media titles. They service equipment in 300 classrooms and run a closed-circuit television (CCTV) system that broadcasts across 26 channels.

MTS has experienced quite a bit of staff turnover in the past few years, the most noticeable being the retirement of its long-time director. As a result, the current management staff has been in place only six months. Figure 4-1 shows a high-level organization chart. Marianne Buchanan was hired from an East Coast university. Her management style is very different from that of her predecessor, Dr. Johnson. Dr. Johnson was a very easygoing boss, more of a friend to the MTS employees. He had been with Desert View for 35 years, serving as MTS director for the last 20 years. His retirement, coupled with university budget reductions, has raised staff insecurity.

MTS is organized into three main units: Production, Instructional Development, and Distribution. *Production* works primarily with faculty and staff in creating educational media such as videos and movies. Mike Guerrero has been the production manager for one year. Prior to his promotion, he worked as a production assistant for four years.

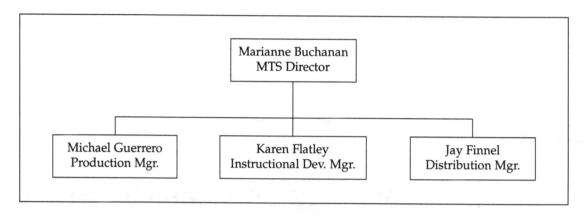

Figure 4-1   MTS organization chart

Instructional Development provides support in creating educational materials. This unit manages a faculty computer lab where faculty and staff have access to a variety of multimedia computers, scanners, and so forth. Additionally, this unit assists faculty in developing course materials. Karen Flatley manages this division. She has been with MTS for only three months.

Distribution is responsible for counter services, closed-circuit television scheduling (CCTV), and media and equipment acquisitions. Counter services provides support in the selection and use of instructional materials and the distribution and maintenance of audio-visual equipment. Counter services staff also check out media and equipment for classroom use. CCTV provides support for in-class projection or transmission of films and videos over the campus closed-circuit system. Media and equipment acquisitions provides support in acquiring new videotapes and films related to teaching, new equipment such as computers and projection equipment, and the maintenance of the on-line libraries. Jay Finnel has been the distribution manager for the past two years.

MTS operates on an academic year calendar, July 1 through June 30. MTS funding is based on the amount of service it provides to the campus. MTS, therefore, needs to keep data for each academic year. These data are used to produce a number of year-end reports and periodic usage records including frequency of material usage, usage by department, and CCTV usage. With the projections for continued budget reductions, it is critical that MTS has access to accurate and timely information. Usage records can be archived at the end of each year. However, customer information and inventory information should be carried forward.

The distribution unit accounts for more than 70 percent of the workload in MTS. The maintenance of the paper-based information in the distribution unit has become problematic. As a result, MTS decided to automate the distribution unit operations three years ago. This automation has resulted in its own set of problems, and MTS has asked that you help them determine some alternatives.

DISTRIBUTION UNIT CURRENT OPERATIONS

Distribution is presently using an unreliable information system called Media-Net. Media-Net started as the pet project of Dr. Johnson. He thought it would be useful to automate some of the paper flows in distribution. The development effort started with a distribution unit employee, Richard Daily, although he only knew a little about databases and programming on Macintosh computers. Richard left MTS about a year after the development started. Since that point, most of the development has been done with student workers tasked with isolated portions of the project. Occasionally, MTS was able to get a programmer on short-term loan from another division on campus. As a result of this inconsistency in programmers, and a lack of overall direction, Media-Net has been in development for several years.

Media-Net is centered around an Oracle 6.07 database, stored on the campus VAX 6300. The user interface is Hypercard running on Macintosh computers. The Macintosh computers are networked using Appletalk. The system is unreliable and error-prone. Examples of problems include not being able to add data, deletion of all of a customer's entries when logging an item return, and inconsistent data such as having the same customer entered with different customer IDs. Because of these problems, MTS staff currently process all transactions in both the paper-intensive manual system and the automated system. Usage of the automated system continues because of hopes that it can be fixed, and because a tremendous amount of data is already entered into the system.

BACKGROUND PROBLEMS

When Marianne Buchanan took over, she recognized that there were many problems with the current operations in the distribution unit. She also identified several background problems that existed in the MTS organization. She believed that these problems contributed to the current state of operations and to the staff turnover the unit was experiencing. The problems are summarized on the following page, along with the current management approach to handling the problems.

1. *Lack of an overall systems development plan.* MTS did not understand the need for systems planning and consequently developed the current system in pieces to fill immediate needs. Ms. Buchanan brings with her many years of experience in successful planning for technology use. Recognizing the lack of significant funding for new technology, she has planned on continuing usage of the VAX 6300. The VAX is maintained by the Campus Computing Center (CCC). CCC will support both Oracle 6.07 and 7. The VAX is connected to the campus Ethernet backbone. In the long run, Ms. Buchanan would like faculty and staff to be able to have on-line access to the reservation system. However, the computing infrastructure will not support this campus-wide at this time.

2. *Lack of any permanent IS staff in the department.* All development was done by temporary IS staff and students. This turnover, coupled with poor documentation, slowed development. While consultants will be used to develop the new system, ongoing maintenance will be performed initially by the Campus Computing Center. Ideally, MTS will be able to hire a full-time IS developer in the next two years. However, until that occurs, expertise will have to be developed within the organization.

3. *Insufficient usage of systems development methodologies.* A thorough systems analysis was never conducted, resulting in incomplete systems specifications. Development was haphazard. Additionally, sufficient time was not allocated to test and document software. As soon as one piece was completed, it was released to the users. Time for documentation was never built into the development effort; as a result it is woefully inadequate. Current managers are hopeful that this new development effort will result in the creation of a workable system. Ms. Buchanan believes that a new system can be operational within one year.

PRELIMINARY ANALYSIS

When Ms. Buchanan took over, she hired a consultant to review the situation. The consultant conducted a preliminary analysis to investigate the options MTS should pursue. Additionally, he led a JAD-like session that identified specific problems with the current system. The results of the preliminary analysis are presented in the consultant's memo, shown in Figure 4-2.

Date: November 10, 1995
To: Marianne Buchanan, MTS Director
From: Ken Griggs, DataSys Consulting
Re: Preliminary Analysis of MTS Media-Net System

During the past month I have investigated the current Media-Net system in use at MTS. I have attached a context-level data flow diagram [Figure 4-3] of your operations, analyzed the file structures of your current database, and worked with your users to identify their needs in an improved system. Overall, the users are satisfied with the intended design and functionality of the Media-Net system; however, they are not satisfied with the current operating status. This dissatisfaction is having a negative impact on employee morale and is resulting in high turnover. Additionally, the staff is concerned about job stability given the recent turnover in management staff. Based on the results of my preliminary analysis, I have identified the following issues that need to be addressed:

1. Response times are unacceptable. Simple queries take as long as five minutes, with reports running an average of 20 minutes. College staff have not been able to accurately identify whether the performance problems are related to the campus network, Hypercard, or the Macintosh computers and Appletalk network.

2. Results are inconsistent. Outputs from queries, report generation, and updates do not give the same results. Additionally, the system crashes frequently.

3. Existing documentation on how to use the system, as well as documentation on the system design, is lacking.

4. The current database is not normalized. Redundancy of data fields and records is widespread, and there are fields that are not used and/or their meaning or intended use is not known.

5. The existing data contains many inconsistencies. With the exception of the on-line media catalog, the remaining data contain too many errors to be salvageable. The amount of time estimated to clean up the data would not be a productive use of time.

6. Hypercard interface is limited. The current system has pushed the capabilities of Hypercard. It is not likely that Hypercard can continue to serve as an interface to the system.

Recommendations:

1. Abandon usage and further consideration of the current Media-Net system.

2. Conduct a complete analysis of your distribution unit.

3. Develop a new system that will allow for efficient operation of the distribution

Figure 4-2   Consultant memo

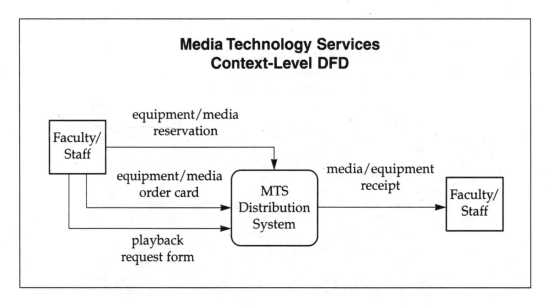

Figure 4-3   Context-level data flow diagram

Based on the recommendations of the consultant, Marianne Buchanan has requested that you conduct a complete analysis of the distribution unit operations. You should present your recommendations for a new system that meet the following criteria:

1. Provide an efficient data model for the storage and retrieval of data.

2. Provide a graphical user interface as the front end.

3. Provide a single entry for all information. This will help reduce the redundancy in operations being performed by counter services. The new system should result in the elimination of the paper system.

4. Provide complete documentation of the analysis and development of the system.

5. Allow for the conversion of the existing on-line media catalog. The current catalog contains more than 7,500 different items, and reentry of this data is not acceptable. Additionally, the system must provide for ongoing maintenance to this catalog.

6. Allow for implementation within a 12-month time period at a cost of no more than $70,000.

DISTRIBUTION UNIT PROCESSES

A customer (faculty or staff) may have several different transactions that are handled through Distribution: 1. equipment/media reservation; 2. equipment/media checkout; 3. equipment/media return; 4. CCTV scheduling; 5. event cancellations. The steps involved in these transactions are detailed below.

Reservations and Checkout

When a request for either an equipment/media reservation or checkout occurs (options 1 or 2), MTS staff will both check the inventory book and do an actual physical inventory check. If the item is available, an order card (Figure 4-4) is completed and the inventory book is updated to reflect that the item has been reserved or checked out. (Reservations are not required for an item checkout, as long as the item is available.) The inventory book contains a listing of all items. Each item contains spaces where reservation/checkout information can be recorded. For reservations, the customer can either schedule the item for delivery to his/her department office or pick up the item.

Items that may be checked out include VCRs, portable computers, video cameras, and VHS cassettes. In the case of a VHS checkout, a cross check with the CCTV schedule must be performed to ensure that the tape is not scheduled to be played on CCTV during the time the user would like to check it out. Most items are identified by either a call number for videos or an ID number for equipment. However, there are some equipment items that do not have ID numbers, yet are still recorded on the order form. These include items like carrying bags, video patch cords, and extension cords.

NAME		ID NUMBER		DEPARTMENT	USER PHONE	DEPT PHONE
COURSE NO.	BLDG/ROOM	CLASS TIME	CHECKOUT DATE	USE DATE	RETURN DATE	
call number	title		accessories		equipment item number	

Figure 4-4   Order card

Returns

When items are returned, the order card is pulled from the file and the return information is completed. The inventory book is updated to reflect the availability of the item. If a user is late three times returning an item, her/his borrowing privileges are suspended for the remainder of the semester. This is recorded on a borrower suspension sheet at the front desk. The sheet is handwritten and contains the name, department, and customer ID for all customers whose privileges are suspended. At the beginning of each semester, a new borrower suspension sheet is started.

CCTV Scheduling

CCTV requests must be made prior to 1:00 P.M. the day before the playback date. Customers may either send in a completed CCTV Playback Request form (Figure 4-5) or phone in the information.

Event Cancellations

When customers call to cancel either a reservation or a CCTV event, MTS must update either the inventory book or the CCTV schedule to reflect the deletion of the appropriate event.

COST AND VOLUME ESTIMATES

Once the system is developed and implemented, the College Computing Center will take over the maintenance of the system. Technical staff in the CCC earn $35,000/year. Personnel in the MTS department average $25,000/year. MTS staff will have the job of training new users on the day-to-day operations of the system. Student workers in the MTS department handle all deliveries and compile all reports. Student workers are paid $6.35/hour. The distribution unit currently processes an average of 85 transactions per day. Table 4-1 shows the breakdown of transactions by number and average time per transaction. The average time stated is for the manual system only. The additional time needed for entry into the current automated system ranges from 4 to 6 minutes per transaction.

The workload is somewhat cyclical, with the average volume increasing by about 12 percent during November, December, April, and May, and decreasing by about 40 percent during June, July, and August.

Desert View Community College
Media Technology Services

CCTV PLAYBACK REQUEST

Requests for playbacks must be received no later than 1:00 p.m. one business day prior to date of playback. Complete and mail this form to Media Technology Services, CCTV scheduling, or place your order by phone at extension 34455.

_____   Circle the Day of the Week   _____
Date of Playback   Mon Tues Wed Thurs Fri   Exact Time of Playback
                                                                       AM or PM

Department _____   Course Number _____

Instructor _____   Building & Room No. _____

Social Security Number _____   Office Phone _____

_____   _____   _____
TV or MP Number   Title of Program   Length

_____   _____   _____
TV or MP Number   Title of Program   Length

_____   _____   _____
TV or MP Number   Title of Program   Length

_____   _____   _____
TV or MP Number   Title of Program   Length

Special Instructions _____

_____

Complete the following only if requesting classroom reassignment:

Class Start Time _____   Class Stop Time _____   No. of Students _____

_____

For MTS use:

Date Received   Taken by   Data Entered   Entered by   Channel

_____   _____   _____   _____   _____

Figure 4-5   CCTV request

Table 4-1   Transaction Information

Transaction Type	Number per Day	Average No. of Minutes
Checkout	25	5
Reservation	12	4
Return	22	3
CCTV Reservation	13	4
Cancellation	5	2

*Note: Transactions do not add up to 85 since some reservations are delivered instead of picked up during a checkout.

REPORT REQUIREMENTS

The system should produce five standard reports, as well as be able to respond to a variety of ad-hoc queries. The first three reports are produced annually, while the last two are produced on a daily basis. Samples of the reports are included in the following figures. All reports are currently generated manually. Estimates of the time required to produce the reports are provided in Table 4-2.

Table 4-2 Time Estimates for Reports

Report Type	Average Time
Equipment Usage	30 hours
Distribution Activity	20 hours
Transactions by Department	50 hours
Overdue Returns	1 hour
Inventory Activity	1 hour

Equipment usage: Produced annually, this report shows the total number of checkouts by item category. Figure 4-6 shows a sample layout of this report.

	Counter Services												
07/01/94	Equipment Usage											Page 1	
Category Description	Jul	Aug	Sep	Oct	Nov	Dec	Jan	Feb	Mar	Apr	May	Jun	Total
TV	99	99	99	99	99	99	99	99	99	99	99	99	999
VCR	99	99	99	99	99	99	99	99	99	99	99	99	999
OHP	99	99	99	99	99	99	99	99	99	99	99	99	999
	999	999	999	999	999	999	999	999	999	999	999	999	9999

Figure 4-6 Equipment usage report layout

Distribution activity: Produced annually, this report tracks the total activity by distribution function. Figure 4-7 shows the sample layout of this report.

07/01/94	Counter Services Activity Report											Page 1	
Activity	Jul	Aug	Sep	Oct	Nov	Dec	Jan	Feb	Mar	Apr	May	Jun	Total
Counter Pickup	99	99	99	99	99	99	99	99	99	99	99	99	999
Delivery	99	99	99	99	99	99	99	99	99	99	99	99	999
CCTV Event	99	99	99	99	99	99	99	99	99	99	99	99	999
	999	999	999	999	999	999	999	999	999	999	999	999	9999

Figure 4-7   Distribution activity report layout

Transactions by department: Produced annually, this report tracks the total items checked out by each department. Figure 4-8 shows the layout of this report.

07/01/94	Counter Services Transactions by Department											Page 1	
Department	Jul	Aug	Sep	Oct	Nov	Dec	Jan	Feb	Mar	Apr	May	Jun	Total
Mech. Eng.	99	99	99	99	99	99	99	99	99	99	99	99	999
Music	99	99	99	99	99	99	99	99	99	99	99	99	999
History	99	99	99	99	99	99	99	99	99	99	99	99	999
	999	999	999	999	999	999	999	999	999	999	999	999	9999

Figure 4-8   Transactions by department report layout

Overdue return report: Produced daily, this report displays all overdue items. Figure 4-9 details this report layout. This information is used by the student workers to produce a late tally sheet. The late tally sheet is a handwritten record of all customers who have a late return. When a customer has her/his first late return, his/her name and ID are written on the list. When the customer gets a second late, a check is added next to his/her name. After a third late, another check is placed on this list, and the customer's information is transferred to the borrower suspension sheet. If the person has more than one item overdue for the same date, MTS only counts it as one late.

			Counter Services			
22-Jun-93			Overdue Return Report			Page 1

User ID Number	User Name	User Phone	Trans Number	Equip Call Num.	Date Due In
123456789	Weiser	555-3421	231446	01273	20-Jun-93
123456789	Weiser	555-3421	231453	01444	25-Jun-93
123111119	Williams	555-3315	231434	11672	15-Jun-93

Figure 4-9   Overdue return report layout

Daily inventory activity: Produced daily, this report is used to indicate all equipment currently checked out. The data is grouped by equipment category. Figure 4-10 details this report layout.

			Counter Services				
30-Jun-93			Daily Inventory Activity				Page 1
			Equipment Checked Out				

Category Description	Equip Number	User ID Number	User Name	User Phone	Date Due In	Status
TV	01273	123556755	Hanson	555-0342	20-Jun-93	Overdue
TV	04233	123456789	Weiser	555-3421	25-Jun-93	
VCR	11672	012311188	Martin	555-3347	15-Jun-93	

Figure 4-10   Daily inventory activity report layout

Media Catalog Information

Figure 4-11 details the information currently stored for the on-line media catalog.

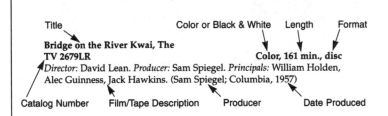

Catalog Number Code Explanations

MP   16mm film
CF   16mm films that are part of the Desert View Library Classic Film Collection
TV   any video format program (videotape or videodisc) that can be transmitted over CCTV
W/G   with guide, libretto, dialogue

The letters following TV numbers must be provided for CCTV scheduling and/or checkout. The letters and meaning are as follows:

X   Beta videotape cassette
XR   restricted Beta videotape cassette
V   VHS videotape cassette
VR   restricted VHS videotape cassette
L   laser disc
LR   restricted laser disc

Note: All non-restricted videotapes, videodiscs, or 16mm films may be checked out.
All restricted programs may be scheduled for CCTV playback or checked out with
prior approval of distribution manager.

Additional Example Entries from the Printed Catalog

Brazil
MP 583   **B/W, 10 min., 16mm**
TV 875   **B/W, 10 min., 3/4″**
People of the Plantation Series, geographical orientation to Brazil, its regional contrasts, coast cities, coffee growing area. (EBF, 1940)

Brickmakers, The
MP 4035   **B/W, 42 min., 16mm**
A picture of exploitation brought upon the poorest group of people in Bolivia today—the families who labor at making bricks. (TRIFL, 1970)

Bride of Frankenstein, The
CF 953   **Reel 1—Sound, B/W, 35 min., 16mm**
                           Reel 2—Sound, B/W, 39 min., 16mm

TV 2696LR   **B/W, 75 min., disc**
Director: James Whale. *Producer:* Carl Laemmle, Jr. *Principals:* Boris Karloff, Colin Clive, Elsa Lanchester. (Universal, 1935)

Figure 4-11   Media catalog information

Exercises

1. Create a data flow diagram to model the current manual processes in the MTS distribution unit.

2. Aside from eliminating the current Media-Net system, do you think the current data flow/business processes need to be improved? Create a new DFD to model your improvements.

3. Create an E-R diagram for the distribution unit.

4. Create a set of normalized relations that accurately model the distribution unit operations.

5. Provide a detailed recommendation of your proposed system.

6. Provide a cost–benefit analysis of your proposed system.

7. Develop the user interface prototype of your proposed system.

8. Given the problems with Media-Net, the MTS staff are very leery of any new information system. This is compounded by their frustration from their current level of overwork. Provide a recommendation to MTS management on how you plan to successfully involve users in the analysis/development of your new system. Create a conversion plan that will help to ensure successful implementation of your new system.

9. Media-Net is able to output the on-line media catalog data in a comma-separated value format. Design the module necessary to convert this into the format needed for your new system.

10. In the future, MTS would like to allow on-line reservation access to faculty/staff. How would this long-term goal affect your design?

Meeting Makers

M eeting Makers provides services to assist organizations or corporations in coordinating and organizing conferences and meetings. Examples of the services provided include handling registration of attendees, fielding questions from attendees, securing meeting spaces and hotel rooms, and planning extracurricular activities. Meeting Makers has been in business since 1988. At that time the only workers were the two owners, Marge Grissela and Stella Shell, and one support staff member. Last year Meeting Makers managed 130 conferences. Over the past six years they have experienced tremendous growth. Overall they have managed the growth fairly well, but they recognize the need for continual improvement. The office is managed by Randy Anderson. He is in charge of the ten project managers, seven office staff personnel, and one graphic designer. Randy reports to Marge and Stella.

When Meeting Makers gets a lead that an organization will be holding a meeting or conference, Marge, Stella, or Randy contacts the organization. Alternatively, some organizations will contact Meeting Makers directly when they decide to hold a meeting or conference. At this point, the client is asked for basic information about the desired event city, dates, anticipated number of attendees, price range, and external activities desired. From this information Meeting Makers prepares a bid for their services. Meeting Makers tries to keep the turnaround time for bids to fewer than five working days. In order to prepare the bid, Randy assigns the project to one of ten project managers. The project managers gather information from the support staff. Depending on the request, they may also solicit information from the Visitors' Center in the desired city.

BID PREPARATION PROCESS

In order to prepare the bid, Meeting Makers staff will gather details and calculate estimates on a variety of services. The project managers are responsible for the overall coordination and compilation of the bid. Unless otherwise specified by the client, project bids typically include all of the following items:

- *Hotel information.* Meeting Maker staff query the existing hotel database (format of database provided in current system section) to identify hotels that meet the size requirements. Then the staff contact hotels to determine availability; and they will provide a note to the project manager with the different hotel options available. The project manager will then review the list and place a tentative hold on certain hotels if the staff believe it necessary (e.g., busy season, not many hotels with large enough space, and so forth). The project manager must remember to call back and cancel if the client decides not to use that hotel. Occasionally, the project manager has forgotten to cancel, and this has resulted in several hotels being unwilling to place holds for Meeting Makers.

- *Registration.* For events with formal registration requirements, such as a conference, Meeting Maker staff calculate the cost of the registration process from a template stored in a registration spreadsheet. Figure 5-1 shows the registration template, including both a data and a formula view. Once the template has been loaded, the projected number of attendees is entered. Based upon that figure, the template calculates the fees for mailing registration information, processing pre-registration, and staffing on-site registration.

  Pre-registration activities include entering attendee information into the database, depositing checks, and generating/mailing a registration confirmation. On-site registration includes staffing the registration booth, registering walk-in attendees, and handling problems that arise at the conference. On-site fees are charged at $250/day per staff person. There is a two-person minimum for up to 100 attendees, with one additional staff member required for every 100 attendees beyond that. The project manager also attends the conference and oversees all staff at the conference. Management service fees are $200/conference day.

  Once the registration spreadsheet is complete, a copy is printed out and sent to the project manager. The completed spreadsheet is not saved electronically. The project manager has the discretion of offering a 10 percent discount on registration fee processing for repeat customers.

- *Brochure design and printing.* The graphic designer provides estimates based upon the number of pages required and the sophistication of the artwork to be included in the brochure. The designer bases his estimate on the type of conference, the number of presentations, and the number of conference activities to determine an estimate of the number of pages in the brochure. The graphic designer does not have this information recorded anywhere, and it is very difficult for Meeting Makers to prepare his bids when he is away from work.

- *Extra activities.* Meeting Makers maintains file folders for each of the major conference cities, as well as other cities in which they have scheduled conferences. Staff members check these folders for brochures that may be of interest to the client. They also scan travel magazines and newspapers. A main calendar of activities is kept in

Meeting Makers	
Registration Worksheet for International Marketing Conference	
Projected Number	
of Attendees	425
Mail Registration	185.94
(.35 * (Attendees +25%))	
Pre-registration	680.00
80% of project # attendees	
(2.00 * 80% value)	
On-site registration	
number of people	5
number of days	4
cost for staff	5000.00
Total Registration Cost	5865.94

a.

b.

Meeting Makers	
Registration Worksheet for International Marketing Conference	
Projected Number	
of Attendees	425
Mail Registration	=B5*1.25*0.35
(.35 * (Attendees +25%))	
Pre-registration	=B5*0.8*2
80% of projected # of attendees	
(2.00 * 80% value)	
On-site registration	
number of people	=TRUNC((B5–100)100+2,0
number of days	4
cost for staff	=B15*250*B16
Total Registration Cost	=B7+B10+B17

Figure 5-1   Registration template: **a.** Data view, **b.** Formula view

the office for the major conference cities. The activities include operas, theater, concerts, attractions, and so forth. Some staff members know of nearby cities that can be cross-referenced for additional activities, but this is not necessarily indicated in the files. When they have gathered the recommendations for activities, they hand the brochures, magazine clippings, or written notes to the project manager.

Once the project manager gathers all of the information, he/she will organize it into a bid. These bids are prepared in a word processing program. Typically Meeting Makers will provide two or three alternatives to the client with a range of costs. Figure 5-2 represents a sample bid detailing one alternative. Once the bid has been provided to the client, the project manager photocopies the information about the extracurricular activities, returns the original brochures to the staff, and places all bid information into a file.

Meeting Makers
99 Broadway
San Francisco, CA
415-555-4444

Supplies and Services Bid for
International Marketing Conference

Project Manager:	Amanda Eklund
Bid Preparation Date:	3/25/96

Conference City:	Dallas, Texas
Conference Dates:	7/1/96 - 7/4/96
Projected Number of Attendees:	425

Mid-Range Bid Information

Hotel Information:	Dallas Regency Hotel Conference Rate: $79 single, $99 double
Activities:	Texas Schoolbook Depository $4/person
Registration Estimate:	$5865.94
Brochure Design/Printing:	$2000.00
Management Services:	$800.00
Total Cost:	**$8665.94**

Figure 5-2   Sample bid

CURRENT SYSTEM

Three years ago Stella and Marge thought they should purchase computers to help the office staff. They read a computer buying guide and visited several retail stores to investigate their options. They purchased twenty 486 IBM-compatible computers, one for each project manager, office staff member, Stella, and Marge. The computers have 8MB RAM and a 200MB hard drive. Each computer has Windows, Excel, Word, and Access. The computer store set up the computers and installed the software and the Novell 3.12 network to share printing resources. The office has two laser printers. Currently there is no sharing of software or files through the network. Meeting Makers contracted with the computer store for technical support when the equipment or software malfunctions.

Meeting Makers has its data stored in an Access system. All conference information is stored in one master database. This database is stored on one PC, and the staff and project managers must alternate usage of the PC to complete their work. Randy designed the database after reading a book on how to create databases. The database contains three tables. The conference table stores general information about a conference or meeting. Data are entered into this table once the organization has accepted the bid from Meeting Makers. The attendee table stores information about people who have registered to attend the conference. This table is created after the conference information and registration forms have been mailed. The hotel table stores information about different hotels. It is periodically updated when staff learn of a new hotel, or when renovations are made to an existing hotel. However, no one staff person assumes the responsibility for the accuracy of the data. The table layouts are shown in Figure 5-3 (primary keys are underlined).

Conference Table

(Conference Name, Organization Name, Address, City, State, Zip, Contact, Contact Phone, Conference Start Date, Conference End Date, Conference City)

Attendee Table

(Attendee Name, Address, City, State, Zip, Phone, Conference Name, Conference Start Date, Member of Organization, Fee Paid, Fee Amount)

Hotel Table

(Hotel Name, Hotel Address, City, State, Zip, Contact Person, Contact Phone, Number of Guest Rooms, Maximum Number of Guests, Number of Meeting Rooms, Maximum Number of Meeting Participants, Largest Meeting Room Size)

Figure 5-3　Current database table layout

Meeting Makers requests that each contracted organization provide an ASCII file of their mailing list. This file contains names and addresses to print mailing labels. These are used to send out information about the upcoming conference. The data is imported into separate Microsoft Word tables (one for each conference) and is used to generate mailing labels.

CURRENT STAFF AND TRAINING

The seven office staff members have all had some basic training on Microsoft Excel and Word. However, they are not very comfortable performing activities outside of their normal scope. Additionally, they have become somewhat self-divided into those who work with Word and those who work with Excel. Only two of the staff members have been trained on how to do queries in Access. For the most part, the current work level keeps them extremely busy and they do not have much free time to spend exploring other capabilities of the systems. The staff turnover averages about one person per year.

MANAGEMENT CONCERNS

Marge and Stella are concerned with Meeting Maker's ability to provide timely and accurate quotes to their customers. Successful meeting organization firms see a repeat customer rate of more than 75 percent. Marge estimates their repeat rate to be 60 percent. She recently asked Randy Anderson to identify some areas where they could improve their operations.

Marge: "Stella, we need to talk. I just got these figures from Randy and am shocked. I can't believe that we lost 15 bids last year due essentially to sloppiness. I just got off the phone talking to our best customer, and they are ready to solicit bids from other companies if we don't get our act together. The bid they got was full of inaccuracies. The hotel we suggested had inadequate facilities, and the quote we gave them for handling registration was based on the wrong number of participants. Randy's memo also indicates that we are not meeting our customers' deadlines."

Stella: "I know. I'm seeing the same problems occurring more frequently as well. I also have seen several penalty charges when project managers have not canceled hotel reservations in time. We need to find a solution. I really can't believe that the data we have can be so bad, and I know we have hired competent workers."

Marge: "I'm really frustrated. We invested a lot of money three years ago when we put computers into the office. I thought the set of programs we purchased would allow the project managers to handle the work. These computers were supposed to eliminate problems. But now I hear from Randy that our database is starting to give him problems. He said it has something to do with primary key integrity. I don't know exactly what he's talking about, but the staff members are starting to alter some of the data to get it to store correctly."

Stella: "Well, it's obvious that we are in over our heads as far as this technology stuff goes. Randy is a great office manager, but I don't think he's going to be able to fix this database problem. Maybe we had better call in a consultant and see if he or she can point us in the right direction."

Exercises

1. Generate a data flow diagram to detail the current processes in place at Meeting Makers.

2. What do you believe is contributing most to the number of lost bids?

3. Do you believe Meeting Makers' current database structure is adequate? Justify why it is, or develop a new structure.

4. Develop a prototype of a system to process bids for meetings.

5. Develop a set of detailed recommendations on how to better maintain the accuracy of data in the hotel database.

6. Marge and Stella are concerned that project managers are not meeting deadlines, such as canceling hotel holds. Suggest a way in which they can better manage this part of the bid process.

7. Given the experience level of the staff and the turnover rate, develop an ongoing training program that will meet their needs.

8. Stella has recently read quite a bit about doing business on the Internet, specifically the World Wide Web. She is wondering whether it would be useful for their company to have a home page on the World Wide Web. Additionally, she thinks they may be able to access useful information about cities and the activities occurring in them. What would you recommend? Justify your recommendation with a cost–benefit analysis.

9. Marge believes the current way Meeting Makers stores brochures and associated material is not the most efficient. Brochures are often not in the files when they are needed, and they are not replaced in a timely manner. She has heard that a scanner can be used to store images of the brochures. Develop a recommendation on whether scanning might be a feasible option to manage this information.

10. Using industry standard figures for computer repair estimates (frequency of breakdowns and costs to repair), provide a cost–benefit analysis relevant to the issue of whether Meeting Makers should look at hiring a full-time computer technician/software expert.

Homeowners of America (HOA)

Homeowners of America (HOA) is a property management firm that primarily manages homeowner associations. Homeowner associations are the entities that maintain the common area in condominiums, townhomes, and many planned unit developments. HOA has been in business since 1986.

There are three partners in the firm. Mike Cadden, the majority owner with 40 percent ownership, oversees all operations and serves as the marketing branch of the organization. He does not like to get involved in the day-to-day details and prefers to be a visionary for the organization. The two other owners, Susan Stone and Art Harvey, each own 30 percent of the company. They split the responsibility for managing the 11 associations. Art and Susan have divided the associations based on size and on the amount of service required for each association. Art manages five associations, and Susan manages the remaining six. Together they provide the overall coordination, ensure that the financial condition of the association is sound, attend board meetings, record and distribute meeting minutes, and handle problems as they arise. HOA employs three office workers (receptionist, bookkeeper, and word processor) and five maintenance workers.

HOA is a fee-for-service organization. The services they offer include the following:

1. Attending board meetings and distributing the minutes
2. Managing financial information
   a. Billing and processing of monthly dues
   b. Paying bills due by the association
3. Performing required maintenance and upkeep on community facilities
4. Communicating with homeowners regarding rule violations

5. Creating and mailing a community newsletter

6. Maintaining records of committee membership

Each service has a fee associated with it. Each association can determine what services they would like. Typically discounts are offered to organizations that contract for more services.

When HOA was created in 1986, there were three associations being managed. Since 1986 the business has increased to 11 associations. The associations vary in membership from 50 to 300 residents. The total number of residents included in all eleven associations is 2,179.

CURRENT INFORMATION SYSTEMS TECHNOLOGY

HOA is using two 486 66MHz IBM-compatible computers. Each system has a 400MB hard drive and 8MB RAM. One of the computers is used by Margaret, the bookkeeper, for maintaining the financial records and committee information. The second is used primarily for desktop publishing and word-processing tasks. The computers are attached to individual laser printers.

ASSOCIATION BILLING PROCESS

Each association's bylaws set the monthly fee, payment period, and late penalties for member dues. Typically these fees are paid monthly, and they generally range from $50 to $200, depending on the services and amenities the association offers and supports. All of HOA's current customers have monthly fees. Seven of the association fees are due on the first of the month, but some are due as late as the 15th. Additionally, the time period after which a payment is considered late varies from 5 to 15 days past due. HOA prints bills and mails them 10 business days prior to the associations' due dates. The seven associations with fees due on the first of the month represent approximately 1,800 customers. This leads to an increase in workload at HOA the last two weeks of the month. Figure 6-1 represents a sample bill.

When dues payments are received at HOA, Rhonda, the receptionist, sorts them by association. Then she manually records the payment amount, date, check number, and customer account number (the county lot number for the property) in a ledger. If a payment does not have a coupon, Rhonda must look at the most recent printout of clients to determine to which association they belong. After the mail has been processed, Rhonda files the checks and payment coupons and gives the ledger to Margaret. Margaret enters the information into an Excel worksheet, such as that shown in Figure 6-2. Each association has a separate worksheet, and data is archived at the end of the fiscal year.

Homeowners of America
P.O. Box 999
Big City, CA 92222
800-555-5555

Community Village 1 Homeowners Association
Monthly Dues Assessment

John Smith
123 Main Street
Anytown, CA 92111
Account 123

Date	Charges and Credits	Amount
03/01/95	March Assessment	99.00
	Total Due.................	99.00

Return this portion with your check made payable to:
Community Village 1

. . . .

Date	Charges and Credits	Amount
03/01/95	March Assessment	99.00
	Total Due.................	99.00

Next Annual Meeting 3/18, 10:00 AM, @ clubhouse
10% Late Charge if not paid by 3/15/95

Figure 6-1   Sample bill

Rhonda is paid $12.50/hour. She estimates that she spends an average of 2 minutes per account, including time spent sorting and looking up missing information. Margaret is paid $18.25/hour. It takes her about one minute per account, including resolving any inaccuracies in the ledger caused by transcription or recording errors.

Community Village 1 Association			1995 Records										
			January			February			March				
Account Number	Name	Address	Date Received	Amount	Check Number	Date Received	Amount	Check Number	Date Received	Amount	Check Number		

Figure 6-2   Association worksheet

DELINQUENCY LETTERS

Payments not received by the end of the late period are considered delinquent. HOA sends out delinquency letters at intervals of 15, 30, 60, and 90 days. Given the varying due dates and late periods for the different associations, Margaret processes delinquencies about twice a week. She uses the autofilter in Excel to determine which homeowners do not have a due date entered for the month. She then copies the account number, name, and address of the delinquent homeowners into a file. She runs a separate filter for each month, and each filter is saved as a separate file.

HOA has a separate disk for each association, and they have adopted the file-naming convention of using the association initials and the month abbreviation. For example, CV1Mar.XLS would be the file name for Community Village 1 Association's March delinquency records. These files are then given to Julie, who runs an MS Word mail merge to print the delinquency letters (see Figure 6-3). She uses a separate Word template to indicate the number of days late depending on the association and the late time period.

Additionally, prior to each board meeting, a list of delinquent accounts is printed and distributed to board members. One problem with this process is that when a customer falls more than one month behind, they will show up in separate monthly files. Mike believes that only one late notice should be sent per customer, regardless of the number of different months that are late. Therefore, Rhonda is responsible for identifying these duplicate letters after they have been printed and consolidating the information into one late notice. She creates a completely new letter that summarizes all of the outstanding late payments. Occasionally, the separate letters do slip by, and a customer may receive more than one late notice.

If a homeowner is more than 90 days delinquent on a special fine, normal bill, or special assessment, HOA files a lien against the property.

Homeowners of America
P.O. Box 999
Big City, CA 92222

January 5, 1995

John Smith
123 Main Street
Anytown, CA 92111

Re: Community Village 1 Homeowners Association

Dear Homeowner,

It has come to our attention that you are 15 days late in the payment of your March
dues. The board of directors has asked that we contact you regarding this matter.

Sections 4.6 and 4.7 of the Community Village 1 CC&Rs detail your responsibility
to make timely payments of your association dues. Section 4.7 specifically details
the procedures the association may follow if you do not immediately make all back
payments owed to the association. It is your responsibility to immediately address
this matter.

Please send in your payment of $108.99:
      $99.00 monthly assessment
       $9.99 10% late fee

Please contact me at 800-555-5555 if you have any questions. Note that failure to
promptly address this matter may result in the board levying additional fines
against you or filing a lien against your property.

Sincerely,

Art Harvey
Association Manager

Figure 6-3   Sample delinquency letter

RULE VIOLATIONS

HOA handles complaints received directly from homeowners or via the board. Complaints typically center around a homeowner violating one of the Covenants, Conditions, and Restrictions (CC&Rs) that apply to their association. Approximately 10 percent of all homeowners are in violation at any point in time. HOA begins the process by investigating the complaint. If they determine the complaint is valid, they will send a notice of violation. The notice asks the homeowner to comply with the rule and warns of penalties if they do not. Figure 6-4 documents a sample complaint letter. Julie (the graphic artist/word processor) is responsible for generating the majority of homeowner correspondence and newsletters.

Homeowners of America
P.O. Box 999
Big City, CA 92222

January 5, 1995

John Smith
123 Main Street
Anytown, CA 92111

Re: Community Village 1 Homeowners Association

Dear Homeowner,

It has come to our attention that you are in violation of the CC&Rs relating to Section 7.13, Landscaping. The board of directors has asked that we contact you regarding this matter.

Section 7.13 requires that all homeowners maintain their landscaping. The board believes that your landscaping has deteriorated to an unsightly and unattractive condition. It is your responsibility to immediately address this matter.

Please contact me at 800-555-5555 when you have resolved the above stated problem. Note that failure to promptly address this matter may result in the board levying a fine against you, or billing you as a result of the board hiring someone to perform the necessary landscaping.

Sincerely,

Art Harvey
Association Manager

Figure 6-4   Sample complaint letter

After 15 days, HOA verifies that the homeowner is in compliance. If he or she is not, a second letter is sent asking the homeowner to attend the monthly board meeting or risk being fined. If the board determines that a fine is warranted, HOA sends out the special notice.

The tracking of letters and necessary followup is done manually by Rhonda. She keeps a paper file (by association) of all letters sent. Each day she checks the files to determine which need followup attention (e.g., send a second letter) and which have been satisfactorily resolved. Because of all the other work in the office, this task often does not get completed (especially when payments are being processed at the beginning of the month). When a complaint has been satisfactorily resolved, the copy of the letter is discarded. No followup correspondence is sent to the homeowner. Problems arise when letters are sent in error because of inaccurate tracking of resolved issues.

SPECIAL ASSESSMENTS

Occasionally, the board may impose a special assessment on the members. This usually occurs as the result of a large, unexpected bill, such as to repair flood damage to common facilities. These assessments are processed in the same manner as bills and are recorded in a special assessment column that is added to the association worksheet.

PAYING ASSOCIATIONS' BILLS

HOA is responsible for paying the bills incurred by the association. These bills may include water, gas, and electricity for any common areas, trash pickup fees, maintenance fees, and HOA fees.

MAINTENANCE PROCEDURES

When an association needs maintenance performed, they will typically request bids from several companies. HOA may decide to bid on a contract. Rhonda is responsible for maintaining routine maintenance calendars for the associations. She schedules HOA employees for their contracts and reminds Art and Susan to periodically check the associations after maintenance has been performed.

COMMUNITY NEWSLETTERS AND COMMITTEE MEMBERSHIP

Julie maintains the committee lists for each association. She keeps this information in a Word document. Each association has a separate document that contains the name of each committee and the names of its members. A printout of the committee membership list for each association is given to Art, Susan, Rhonda, and Margaret. This allows them to answer questions when a customer calls. When an association contracts for a community newsletter,

Julie is responsible for creating and printing the newsletter. She really enjoys this part of her job, since it allows her to use her graphic design training.

ADDITIONAL INFORMATION

Mr. Cadden would like to be able to add another 10 organizations in the next five years. However, he is concerned that HOA's growth may be limited due to the inefficiency of the office procedures. The staff are not able to handle any additional work. He has also noticed an increase in errors (such as improper crediting of accounts or inaccurate violation letters) during the last two weeks of the month. The resulting rework is further complicating the situation. After discussing the situation with Art and Susan, Mr. Cadden has come to the conclusion that simply hiring additional office staff will not alleviate the problems.

HOA would like you to provide an analysis of their current operations and make recommendations on how they can best improve their current procedures. Ideally they would like to automate as much as possible, provided they can justify the costs. They would also like to see a prototype of your recommended solution. Mr. Cadden would like a comprehensive proposal that addresses his current needs but will also allow him to manage comfortably for the next five years.

Exercises

1. Provide a data flow diagram of the current activities performed in the office.

2. Identify any manual activities that you think can be re-engineered.

3. Provide specific recommendations for re-engineering these manual activities.

4. Do you think the current design being used by Margaret (separate worksheets for each account) will continue to be satisfactory, given the growth projections? Provide specific recommendations on how to improve this process. Prototype your solution. Your solution must prevent Margaret from getting duplicates when she performs the query for delinquent accounts.

5. Currently, Rhonda must check through her file of letter copies she has sent out regarding violations and delinquencies. She must remember to check and see if the problem has been resolved. Suggest an alternative way to handle this situation.

6. Taking into account your prototype solution, provide a set of recommendations for upgrading the existing hardware/software. Ideally this upgrade will last five years.

Hazardous Materials Management System (HMMS)

Huntington County is a mid-size county in the southeastern United States. The Public Works Department (PWD) of Huntington County has recently assumed responsibility for ensuring that the county meets all current and future safety and regulatory requirements for hazardous materials. Hazardous materials the county must track include containers of chemicals, paints, cleaners, and so forth. The types of information that must be tracked include the concentration of chemicals, type and size of storage container, storage location, and hazard control method.

Prior to the reorganization, all individual county departments were responsible for their own reporting. New regulations set to take effect next year require that all government and private entities establish new, tougher reporting requirements. The county hopes that this reorganization of responsibilities will help ensure its compliance with the federal regulations. Tom Rogers, the manager of the PWD, recognizes that technology may provide certain advantages to help him comply with the regulations. The county board of supervisors agreed with Tom and suggested that he contact the Information Systems Division. Tom outlined the primary scope and objectives for this project in his memo (Figure 7-1) to the county Information Systems Division (ISD).

Date: January 5, 1995
To: Michael Harvey
         Information Systems Division Manager

From: Tom Rogers
         Public Works Department Manager

Re: Development of Hazardous Material Management System (HMMS)

After meeting with the county board of supervisors and surveying the current situation in several county departments, I have created the primary scope and objectives for the HMMS.

- Monitoring and reporting of material inventories to meet safety and basic regulatory requirements
- Tracking and reduction of hazardous waste to reduce hazards and costs associated with disposal, and to meet regulatory requirements
- Management of excess inventories to reduce hazards, decrease liability, and reduce stocks of materials that spoil with age
- Management of hazardous materials storage by hazard type, per regulations
- Development, dissemination, and implementation of safe hazardous-materials handling procedures
- Distribution of material safety information (e.g., Material Safety Data Sheets [MSDSs])
- Planning for new facilities to meet material safety requirements
- Support for a uniform safety training program
- Providing accurate and timely information to emergency response personnel

Please review the feasibility of this request and schedule a meeting with me at your earliest convenience.

Figure 7-1   Memo to Information Systems Division

After receiving the memo, Michael decided to perform a preliminary analysis to aid in determining the feasibility of this new system. During the analysis he took the following notes:

- The basic processes performed on hazardous materials include ordering, receiving, storing, dispensing, and disposing.

- There are more than 30 different departments/areas that currently have hazardous material. The data storage in these departments ranges from manual systems to simple databases. There is no consistency across departments as to the data fields captured or the formats of the fields.

- It is estimated that more than 50,000 different containers of hazardous materials exist in the county. The county orders approximately 30,000 containers a year.

- The state requires that MSDSs be manufacturer-specific; therefore, generic MSDSs cannot be used. However, there are companies that provide an electronic format for MSDSs.

Michael's next step was to schedule a meeting with Tom. At the meeting he hoped to clarify the scope and present the information he discovered and the plan he had for attacking the development.

Tom: "Michael, thanks for such a quick response to my memo. As you can see, we are really under the gun if we want to meet the government requirements before their scheduled inspection in November."

Michael: "Well, this is really going to be a challenge. We are going to have to reorder our priorities in ISD, and that's going to upset a lot of other county users who were counting on new systems."

Tom: "The board can smooth over any ruffled feathers."

Michael: "Yes, but our biggest problem will be getting all of the various departments on board. Everyone is doing their own thing right now. If this new HMMS is going to be successful, the county must implement logistical, procedural, and policy changes. With the history our county has of decentralized responsibility, this shift back to centralization will not go over smoothly."

Tom: "Don't worry about that. If they want to avoid regulatory penalties, they will change. So what do you think this new system should do, and what's it going to cost?"

Michael: "Well, in addition to the basics that we both know about, like storing the size and location of each container, the system should track storage procedures so that users can easily find out whether chemicals can be stored next to each other. One of my initial concerns is determining the best method of tracking the containers. Right now it is pretty easy for each department to just look at what is on its shelves. But the computer will need to be able to ID each container. After ISD has completed the requirements definition for the new system, we will be able to propose some alternatives."

Tom: "That sounds good, but make sure we can afford your proposal. I also thought of something else that needs to be addressed. Ever since you started asking questions about the different ways the departments track hazardous materials, I have been getting calls from people screaming at me about who will be able to see their data. You are going to need to prevent people from looking at the chemicals stored in different departments."

Michael: "We can certainly investigate security concerns, but getting back to your cost concerns, I will need some more data before I can give you an estimate. Based on my preliminary estimates of system size, and the fact that many departments have no computers or very old ones, I think we are looking at a significant investment in new hardware. Because of the budget limitations for this year, realistically I think we will need to phase in the development of the system. We can probably get a prototype up and running for the PWD by November. Hopefully this will be enough to satisfy the regulators. Once the final system has been completed, I imagine we will phase it into the other departments as the budget allows."

Tom: "Keep me posted, and I will let the board know about your thoughts."

When Michael got back to his office, he knew that this would be an ambitious undertaking for ISD. Given the current hardware and software configurations of the PWD, and his preliminary ideas about the needs of the new system, his initial thinking was that they would be best served by a client–server-based system. He called together his best developers and put them to work immediately.

Wyatt Tyler started with a detailed analysis. He conducted staff interviews, gathered forms used in the different departments, toured the different facilities, and did a preliminary data analysis. One of the forms he found was a diagram that showed the basic flow of data (Figure 7-2). He was unable to find out who created the diagram, and although it didn't use standard notation, it did accurately describe the basic flows.

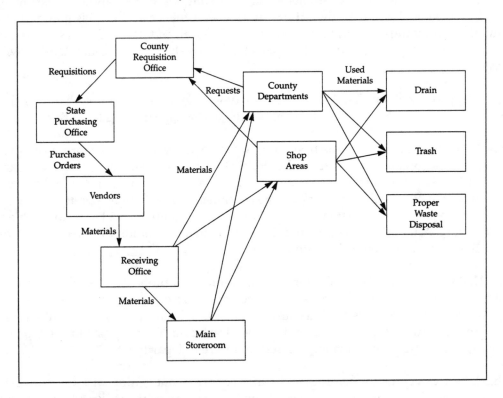

Figure 7-2   Basic flow of data

The process starts when the departments or shop areas generate a request for materials and ends with either the usage of the material in the departments or shop areas or the disposal of the used materials. Wyatt also found the inventory form presently used by the departments to gather data about hazardous materials. An example of this form is shown in Figure 7-3.

A physical inventory count is performed three times a year. The inventory forms are maintained in individual departments, and they produce their own summary reports about hazardous materials. The departments do not routinely track the exact usage or disposal of materials. As a result, for any one container, it is not possible to report on how much was used, what was disposed of, and by what method.

After analyzing the information he had collected, Wyatt recommended that it would be necessary to track hazardous materials by container. He also suggested that the easiest way to be able to monitor the containers would be through a bar-coding system. The receiving office would be responsible for tagging each incoming container. When materials are sent to a department, the department would scan the container and update the location on the record. Each time a material was used, the record for the container would be updated to reflect its usage and eventual disposal.

Hazardous Materials Inventory
Actual Inventory

Inventory Date: _____   Department: _____
Location/Bldg: _____ Room: _____   Description (lab, storage, etc.): _____
Supervisor(s): _____   Phone: _____
Person(s) performing inventory: _____   Phone: _____

This information is requested for every hazardous chemical in your area of responsibility. Please complete this form to the best of your ability.

EHS Code #	Chemical Name/s or Tradename	Manufacturer or Supplier	CAS #	Form (circle one)	Qty.	Units	MSDSs on Hand	Hazard Class
_____	_____	_____	____	Sol Liq Gas	____	____	Y N	____
_____	_____	_____	____	Sol Liq Gas	____	____	Y N	____
_____	_____	_____	____	Sol Liq Gas	____	____	Y N	____
_____	_____	_____	____	Sol Liq Gas	____	____	Y N	____
_____	_____	_____	____	Sol Liq Gas	____	____	Y N	____
_____	_____	_____	____	Sol Liq Gas	____	____	Y N	____
_____	_____	_____	____	Sol Liq Gas	____	____	Y N	____
_____	_____	_____	____	Sol Liq Gas	____	____	Y N	____
_____	_____	_____	____	Sol Liq Gas	____	____	Y N	____

Notes: EHS Code # is a seven-digit numeric identifier. It is a unique number that has been assigned by the county for a particular type of chemical. There should be a one-to-one correspondence between the EHS Code # and the CAS #.

CAS #: Chemical Abstracts Services #. This is an on-line service that assigns identifier numbers to chemicals.

Hazard classes should be listed according to the Aldrich Chemical Catalog.

Figure 7-3   Hazardous materials inventory form

Michael began identifying system constraints. He started by looking at constraints surrounding equipment availability. The county uses a variety of Macs and PCs. They have no plans to standardize a particular model; therefore, any system would need to be able to run on both platforms. Most of the existing units would need to be upgraded. The current systems typically were 286-based machines with less than 4MB of RAM. The county file-server has enough space and processing capability to handle the new system. The county data network would adequately handle network traffic, and dial-in access could be supplied where needed. Bar-code equipment would need to be purchased. Based on his investigation into possible options, Michael recommended that the following minimum configuration be met in each of the departments requiring access to the HMMS system:

Basic PC Platform:

> 486 or better. Minimum of 8MB RAM.
> Hard-drive space: At least 500MB.
> Operating Systems: Windows 3.1 or Windows 95.
> Display: 800 × 600 or better
> Network capability: Ethernet connection

Basic Mac Platform:

> 68040 or PowerPC. Minimum 8MB RAM.
> Hard-drive space: At least 500MB.
> Operating System: System 7.5 or better.
> Display: 14-inch or better color monitor.
> Network capability: Ethernet connection

From Wyatt's interviews and observations of the different users, he learned that the typical users of the system would not be computer literate and that the application would only be used during about 10 percent of their workday. Based on this, he recommended that the new system be developed with an easy-to-use graphical user interface. Additionally, the system would need to contain a fairly detailed help system.

Since many other organizations also are required to track hazardous materials, Michael thought there might be off-the-shelf software available. He assigned Suzi Orga the task of identifying off-the-shelf products. After Suzi received both Wyatt's recommendations on the specifications for the new system and the constraints identified by Michael, she was able to present the results of her research. Her results are presented in Figure 7-4.

To: Michael Harvey
From: Suzi Orga
Re: Summary of Commercial Alternatives for HMMS

I investigated approximately 16 different commercial packages that could possibly meet our needs. After comparing them to the requirements set forth by Wyatt and you, I realized that none would meet our requirement for a client–server-based system. However, many of them had a good prospect of evolving in a client–server direction, so I continued to consider them. The results of my analysis of those systems are summarized below:

- None would meet all of our needs. Some handle MSDSs but not material tracking, or vice versa.
- The end-user software was not available for both Macs and PCs.
- The host machine is a platform that is not available to the PWD.
- Bar-coding was not supported.
- User interface was poorly designed.
- The tasks to be managed did not match our needs.

Based upon my investigation, I believe that we will need to develop our own system. If you would like any more details on the packages I surveyed, please let me know.

Figure 7-4   Memo from Suzi Orga

After reviewing all of the material his staff gathered, his notes from the meeting with Tom, and the system request from Tom, Michael finalized the minimum requirements of the new system. These requirements are as follows:

- Capture and manage data about hazardous material inventory, including receiving, usage, disposal, and so forth.

- Make available MSDSs and related safety information to material handlers and users.

- Manage information about the identity of authorized material users.

- Manage information about locations, including safety facilities at each location.

Michael reviewed these minimum requirements with Tom. At the same time, Michael presented his development plan. He thought it would be best to develop the system in two phases.

In the initial phase, ISD would create a static inventory system that would allow for tracking by container and would include the bar-coding of all hazardous materials. A centralized database will be created to store all of the needed information on a container basis.

During this time new procedures could be created, and the additional time could be used to build users' support from the various departments, as well as sort out the budget problems. Additionally, current usable data could be converted and reports could be created to meet the government reporting requirements.

This system would be housed in the Receiving Department. As orders came in, the staff in receiving would bar-code the containers and enter the appropriate information into the system. The basic information would relate to the hazardous material, vendors, location of the material, and whether an MSDS had been entered into the system. Physical inventory checks would still be conducted by all departments using forms similar to that shown in Figure 7-3. The main report that the system should provide for government purposes would need to identify each hazardous material that was purchased and show how it was used and disposed of. It would also summarize the quantity of material disposed of by each of the main options (drain, trash, proper disposal).

The second phase would involve the movement to an interactive, integrated management and control system. This would add features that could handle the disbursement of materials and interact with purchasing, receiving, disposal, and so forth. Additionally, this system should have an up-to-date status on the location of any material, including how materials were used and disposed of. Standard requisition and purchase order forms could be integrated during this phase. Phase 2 would also involve all end users having access to the system. This would allow them to update the system on an as-needed basis when different materials were used and/or disposed of.

Exercises

1. Memos from users often have problems associated with them. For example, how can the HMMS "reduce hazards associated with disposal"? Convert the memo from Tom Rogers into a scope document of what an HMMS information system can really do.

2. Convert the data flow diagram that Wyatt found into one with more standard notation.

3. Create an E-R diagram for the HMMS system.

4. Create a set of normalized relations to handle the requirements for the Phase 1 system.

5. Modify the relations created in question 4 to handle the added requirements from Phase 2.

6. Given the fact that this system will involve changing from a decentralized environment back to a more centralized system, how can the users be encouraged to convert to this new method?

7. Using a standard project management technique, develop a plan for how to accomplish the development of this system through Phase 2.

8. Create a set of recommendations on handling the transition of users from their current system to the Phase 1 system and then to the Phase 2 system.

9. Develop a prototype of the Phase 1 system.

10. When the Phase 2 system is complete, anytime hazardous material is moved or used it will have to be entered into the system. Develop a set of procedures for the entry of this type of information into the Phase 2 system.

National Booksellers, Inc. (NBI)

National Booksellers, Inc. (NBI) began operations in 1990 as a magazine distribution operation. NBI now buys books and compact discs from publishers and large distributors and resells them to bookstores, computer stores, and other retailers. NBI, headquartered in California with a distribution center in New York, is currently the smallest company of a six-company enterprise owned by Bill Tyler. However, NBI's growth opportunities appear unlimited if the organization can develop adequate systems to support the business. As an example, a large retail chain has agreed to purchase books from NBI provided that they can have a new information system on-line within six months.

Given these opportunities, NBI was chosen as the first company of the enterprise to convert to a new information system. Eventually, all six subsidiaries are planning to install new systems that should, if properly developed, leverage the enterprises' resources (e.g., people, procedures, data, software, and hardware). This goal is clear to management and will certainly be considered when information systems decisions are made for NBI.

COMPANY ORGANIZATION

The key people in the NBI organization are those people who have the responsibility and authority for all six companies within the enterprise.

Bill Tyler is the CEO and sole proprietor of the six-company enterprise. His management style is autocratic, and he completely controls all decision-making. For example, virtually every payable must be personally signed by him, and he often forwards invoices to department managers requesting detailed explanations of charges.

Marie Corbitt, vice president, is primarily responsible for all marketing decisions. Her style is casual and relaxed. She started the parent company with Bill 12 years ago and advises him on company-related issues and problems, especially if the outcomes affect the strategic direction of the company.

Richard Gonzales, director of marketing, is a two-year employee who is primarily responsible for acquiring new publishers and retailers. He is also responsible for maintaining adequate inventory levels. Richard has no background in information systems, and he relies on the IS staff for most of his information systems needs.

Carol Bailey, accounting manager, has been with the company for five years and is responsible for payroll, accounts receivable, and human resources for each of the six businesses. The company has no controller, so the accounting manager must also prepare preliminary financial statements each month. The monthly financial statements are then forwarded to an outside accounting firm.

Michael Nakatani, manager of warehouse operations, is located in New York, where he oversees all of the book shipments and book returns. Michael has been with the company since NBI acquired the warehouse in New York in 1991.

John Lackritz, operations/IS manager, has been with the company for eight years and has been in his current position for three years. He is responsible for the development of the new information system. The tasks that he will oversee for the project include writing the specifications for the new software, signing off programming changes, approving purchases for software and hardware related to the project, creating and maintaining the project schedule, training, writing new procedures, system documentation, producing regular progress reports, and transitioning to the new system from the old. He will continue to oversee the operations of the existing system while he manages the development of the new system.

With the exception of Michael Nakatani, management and all of the approximately 50 accounting, marketing, and customer service employees live in California. Another 50 NBI employees work in warehouse operations in the New York area. Figure 8-1 depicts NBI's management structure.

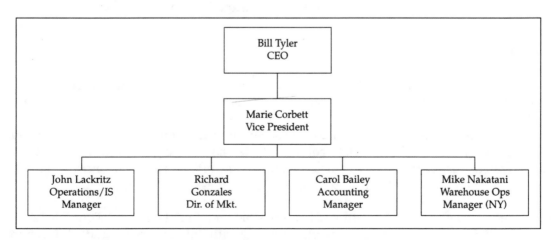

Figure 8-1   NBI's management structure

BUSINESS AT NBI

NBI buys books and compact discs from publishers and large distributors and resells them to its customers: bookstores, computer stores, and other types of retailers. The business operations at NBI involve customers, order entry, publishers/vendors, inventory, invoicing, shipping, and returns. A description of how these operations are performed is given in the following pages.

Customers

Pertinent data about NBI customers is entered on the NBI customer information form shown in Figure 8-2. This and most of the other data entry forms are screen forms with paper backups in the event the NBI information system goes down. The same form is used both to gather data about new customers and to make changes to existing customer information. NBI currently maintains data on 136 customers. They anticipate that the number could double within the next three years.

NBI Customer Information
Address Change/Status Change

NEW/CHANGE   DATE _____

Ship To Address

Cust No	Phone
Company	Fax
Address	Contact
	Title

| City/State | Zip | Country |

Bill To Address

| Company |
| Address |

| City/State | Zip | Country |

| Resale #: | Sales Discount %: |
| Type (Below) | Industry (Below) | Mkt Code (Below) |

Type		Industry		Mkt Code	
C	Chain Store	BK	Bookstore	0	Other
FR	Franchise	CP	Computer Store	1	Wholesalers
P	Publisher	NE	Newsstand	3	Barnes & Noble
R	Retailer	OT	Other Industry	4	Bookstores
W	Wholesaler	PB	Publisher	7	Software Stores
		WH	Wholesale	9	Tower Books

Figure 8-2   Customer information form

Order Entry

An order taken at NBI may be a first-time order, a backorder, and/or a reorder. During the order process NBI's customer service staff will enter a customer number and then one or many line items for the order. The operator must look for restrictions that prohibit NBI from selling certain items to certain types of customers. For example, another distributor may have exclusive rights to sell to a particular retail chain, prohibiting NBI from selling an item to that retail chain. Since multiple codes can be entered in the market code field in the inventory form, the operator must manually check for restrictions when entering a line item on sales to that customer. It is important that customer service not be allowed to enter restricted items when the order is placed. Once the order is complete, a quantity discount may be applied to the order. Figure 8-3 shows NBI's sales order form.

Publishers/Vendors

NBI purchases books, magazines, and CDs from various publishers and vendors. NBI maintains data about these publishers/vendors, including payment terms and return policy information. The screen form shown in Figure 8-4 is used to collect this information. Currently NBI deals with 15 different publishers/vendors.

Date		Customer #		Customer Name			
Item #	Qty	Title		Cover Price	Disc Price	Extended Price	
1.							
2.							
3.							
4.							
Total Qty					Sub-Total		
					Quantity Discount (–)		
					Shipping (+)		
					Balance		

Figure 8-3   Sales order form

Publisher/Vendor Information		
Payment Information		
Company Name	Vendor ID #	
Address		
City	State	Zip
Phone #	Fax #	
Contact Name	Title	
Payment Terms		
Returns Information		
Company Name		
Address		
City	State	Zip
Phone #	Fax #	
Contact Name		

Figure 8-4   Publisher/vendor information form

Figure 8-5 shows a simple entity-relationship diagram for order entry data.

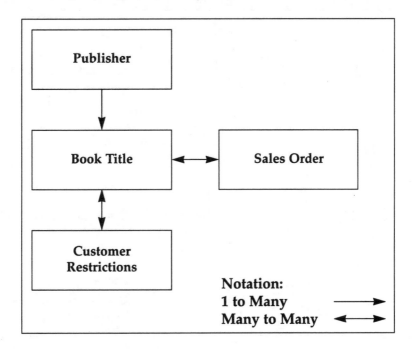

Figure 8-5   Order entry E-R diagram

Invoicing and Shipping

Invoicing and shipping are tightly integrated. Weekly scheduled shipments are delivered with invoices to NBI's customers. To generate an invoice (Figure 8-6), all freight charges must be calculated to be included on the invoice. The process of determining the shipping method, shipping cost, and shipper requirements is a very complicated one and has been automated for the other enterprise companies and will be integrated into the new software at a future date.

Inventory

Inventory consists of books, magazines, and compact discs. A sample inventory form is shown in Figure 8-7. The market code is used to indicate customer authorizations/restrictions on the material. For example, an item with market codes 2 and 9 indicates the item can only be sold to Barnes & Noble and Tower Books. In addition to the information on this form, the quantity on hand, quantity backordered, and quantity on order must be tracked in the new system. Inventory is adjusted when books are invoiced or credited to customers or returned to vendors, and when physical counts are completed. The current inventory consists of approximately 1,200 different items, including 850 books, 120 magazines, and 230 CDs.

NBI Invoice					
Cust No:					DATE _____
		Ship To Address			
Company		Phone			
		Fax			
Address		Contact			
		Title			
City/State			Zip		Country
		Bill To Address			
Company					
Address					
City/State			Zip		Country
Item #	Qty	Title	Cover Price	Disc Price	Extended Price
1.					
2.					
3.					
4.					
Total Qty				Sub-Total	
				Quantity Discount (–)	
				Shipping (+)	
				Balance	

Figure 8-6 Invoice form

NBI Inventory Form		
Date		
Item #:		
Title:		
UPC #:	**ISBN:**	**Vendor #:**
Product Line (Below):		
Pricing:		
U.S. Price:	**2nd Price (Dealer)**	**Disc %**
Published Date:	**Publisher Name:**	
Market Codes (Below):		
Author:		
Disc Code (Circle One):	**CD - W/Compact Disc**	**WD - W/Diskette**
Return Code (Circle One):	Full/Covers/Affidavit/Masthead/Non-Returnable (Circle One)	

Product Line		Market Code	
B	Book	0	Other
C	Compact Disc	1	Wholesalers
M	Magazine	2	Barnes & Noble
		3	Computer Stores
		4	Bookstores
		7	Software Stores
		9	Tower Books

Figure 8-7   Sample inventory form

Customer/Vendor Returns

All titles purchased by NBI's retailers can be returned for credit at 50 percent of the retail cover price. Returns can be made by retailers after 90 days from receipt without prior notice and must be accompanied by the return form. Books proclaimed *called-in* must be returned (freight prepaid by the retailer) within 60 days from receipt of notice and be in salable condition (free of stickers and markings). NBI is not responsible for lost returns. Returns are shipped via United Parcel Service (UPS), Roadway Packaging Systems (RPS), or traceable mail to the distribution center in New York.

Customer return data includes the customer number, a return number, return date, book item number, quantity returned, return discount, and extended totals for each book returned. NBI returns unsold books and books returned from its customers to the publisher/vendor.

NBI uses the same type of information it collects on returns for its publisher/vendor returns. Publisher/vendor returns contain a vendor number, return reference number, return date, book item number, quantity returned, book price, return discount, and extended totals for each book returned.

Customer Statements

At the end of each month, customers are mailed a customer statement, shown in Figure 8-8. The customer statement shows a summary of the month's transactions. It includes the customer number, customer name, address information, beginning balance for the month, ending balance, statement date, customer accounting transactions that occurred during the period—such as charges, credits, and payments—and an ending balance.

EXISTING INFORMATION SYSTEM

The current NBI information system consists of dumb terminals that are used for processing orders and other customer service activities. These terminals are connected to a set of Motorola 6000-based microcomputers running a proprietary operating system called AlphaMicro. The order entry software that runs on this system is also proprietary and is completely dependent on the AlphaMicro system. NBI has one full-time AlphaBasic programmer who maintains these systems. Additionally, NBI has an off-the-shelf accounting package running on one of its three stand-alone, DOS-based microcomputers. A word processing package and a spreadsheet are the primary applications installed on the other two machines.

Customer Statement			
Account No.:		Date:	
Current	30 Days	60 Days	90 Days
xxxxxxx	xxxxxxx	xxxxxxx	xxxxxxx
Date:	Order/Return #:	Description:	Charge Credit:
	Balance Forward	xxxxxxx	
	Total Charge	xxxxxxx	
	Total Credit	xxxxxxx	
	Balance Due	xxxxxxx	
Customer Name:			
Address:			
City, State, Zip:			

Figure 8-8   Customer statement

John Lackritz feels NBI has outgrown the capabilities of the current software and hardware platform, and the company is reluctant to continue investing in outdated, proprietary technology. John summarized the major problems with the current system in the memo to Bill Tyler shown in Figure 8-9.

To: Bill Tyler

From: John Lackritz

Re: Development of New Information System

As you are aware, we have the capability to attract several additional NBI clients if we can develop a new information system within the next four months. Because of the limited staff available within the IS division, I am recommending that we hire a consulting group to work with us on this project.

As a first step in this project, I have identified some of the major problems with the current system, as well as the high-level requirements of the new system.

Problems with Current System

- The vendor of the current proprietary system offers limited support.

- Upgrades to the operating system are not frequent enough to keep up with the needs of the company.

- System lacks adequate ad-hoc reporting features.

- System has become very slow and unstable.

- System performance is inadequate for current number of users; number of users is expected to increase.

Requirements for the New System

- Software applications should be able to run on virtually any hardware platform.

- The major portion of the new application should be on-line within four months.

- The base package must be affordable.

- NBI must have rights to own and modify the program code.

- Data must be exportable to existing accounting package until conversion to new package is successful.

With your approval, I would like to hire the consultants immediately and begin the detailed analysis and development for the new system.

Figure 8-9   Memo to Bill Tyler

PROJECT DESCRIPTION

John received Bill's approval last week. As a result, you have been hired to provide expert assistance in the analysis and development of the new system. In your first meeting with John, he shared with you his vision of the NBI project. He indicated that the company has become increasingly frustrated with the performance and difficulty of maintaining the current software. He envisions that the new system will provide more flexibility, speed, and control over what is produced by the computer system.

There are two sites to be considered in the initial design of the new system: California and New York. Currently, there are 100 users on the system—50 in California and 50 in New York. The two sites will have to communicate with each other. New York handles all shipments and returns, while California handles accounting, marketing, and customer service functions. It is estimated that 15 of the California users and 7 of the New York users will be on the system approximately 80 percent of their workday. Of the remaining 78 users, it is estimated that their usage will average approximately 25 percent of their workday. Additionally, both locations have extensive printing requirements. California prints monthly statements and most reports; New York prints all shipping documents and bar-code labels.

Exercises

John has asked that you provide him with assistance by completing the following tasks. When appropriate, remember that you need to consider both the long-term needs of the entire organization and the shorter-term needs of NBI.

1. Given Bill Tyler's management style, what type of analysis/development methodology would have the most likelihood of success?

2. Create a context diagram and a level 1 DFD (data flow diagram) that would illustrate the high-level processes at National Booksellers, Inc.

3. Complete the E-R (entity-relationship) diagram for the NBI information system.

4. Based on the E-R diagram created, develop a set of normalized relations.

5. How do two sites (East and West Coast) present problems with system development? How might these problems be overcome?

6. Taking into account that this company is understaffed and the staff does not have extra time to work with you, provide a short report to John indicating your recommendations for an effective way for you to elicit system requirements.

7. Provide a cost–benefit analysis of the system you are proposing.

8. Develop a system prototype.

9. What recommendations would you make to ensure that the system developed for NBI will fit with the system being developed for the overall organization?

TechPrint, Inc.

TechPrint, Inc. is a 10-year-old company that specializes in the printing of hardware and software documentation for high-tech firms in the Silicon Valley area. The firm has approximately 50 employees and is located in the East Bay, about an hour from San Francisco. Chad Notly is TechPrint's president and the principal owner of the company.

Chad is very aware that technology is changing (or will soon change) the way business is conducted; he reads about such things as the Internet and the World Wide Web, and he hears about promising new technologies such as Java and search agents from his high-tech customers. In fact, Chad has become very enamored with this type of technology. For example, TechPrint has become a big user of Federal Express since it developed and distributed software that allows TechPrint (and anyone else with a computer and modem) to print shipping labels, request a courier, and track packages with just a few keystrokes from a computer.

There is another side to this type of technology, however, that is currently unsettling to Chad. Very soon, he suspects, his clients will *expect* to deal with TechPrint, and their other suppliers, in the same way that he now deals with FedEx—using state-of-the-art technology. In anticipation of these changes, Chad is considering ways to bring this type of technology to TechPrint.

The technology Chad is most interested in for TechPrint is that dealing with the World Wide Web (WWW). He recently read in *Business Week* (December 4, 1995, page 83) that there are Web pages that allow people to shop on-line simply by dragging pictures of items into an icon of a shopping cart. To Chad, this Internet version of the FedEx model is akin to what he wants to do at TechPrint, which is to allow his customers to order hardware and software documentation easily over the Web.

THE PROBLEM

Chad would like to hire you to help him plan the integration of this type of technology into TechPrint. He does not expect you to actually develop a home page for TechPrint with the drag-and-drop functionality previously described, but he does want you to help TechPrint develop a plan for use of the Web, including a comprehensive feasibility (operational, technical, and economic) analysis.

In other words, Chad would like to know what it will take for TechPrint to successfully acquire, use, and maintain this type of technology. He realizes that currently there are a number of alternatives that must be considered before TechPrint can adopt this technology. As an example, he knows that he can hire an Internet access provider to develop and maintain his Web page at its site, or he can have someone develop it and then his staff can maintain it at TechPrint. He has heard about ISDN and T-1 lines but doesn't know what these technologies really do, or if/how he can use them to service his customers.

Chad doesn't have any details or opinions on these issues. He is also afraid of the issues that he doesn't yet know anything about. This is where you can be invaluable to TechPrint. Chad hopes that your analysis and plan will help him make a better decision before TechPrint adopts this technology.

From your initial discussion with Chad, you can assume that he wants TechPrint not only to be one of the first printing firms to have a World Wide Web presence, but for TechPrint's presence to be one of the most functional in the industry. This means that his decision is not *whether* TechPrint will adopt Web technology, but rather *how* TechPrint will acquire this technology. You know up front that the decision about adopting the technology will not hinge on an economic feasibility analysis; Chad views TechPrint's use of the Web as a strategic decision and expects that this technology will cost more than it's worth, at least for the first year. It is important, however, that Chad be made aware of the expected costs (such as the cost to maintain a Web server in-house) and given some idea of the tangible and intangible benefits (such as estimates of revenue growth) that can accrue from this technology.

CURRENT TECHNOLOGY AT TECHPRINT

Chad and most of the employees at TechPrint are technology-oriented and are not shy of acquiring new technology if they believe it will help them do their jobs more productively. Currently, all of the computers at TechPrint are Pentium-based PCs running Windows 95 and networked by a Pentium 120MHz running Microsoft's NT server software. Chad would like to grow his business on a single, integrated technology platform such as he has with Windows 95 and NT rather than acquire, for example, a UNIX-based machine if the Web project comes in-house. This, he believes, will lessen the need to hire more than one IS specialist.

The current computer configuration at TechPrint was set up by a consultant who will be leaving TechPrint at the end of the quarter for personal reasons. Chad is interviewing now to fill the position with a competent, full-time IS person who will come on board about the same time. However, Chad does not want to wait until that person is actually hired to begin the Web effort. The new hire should be sufficiently competent to maintain the NT network and whatever IS technology is acquired to run with this system.

Exercise

Assume you have agreed to undertake the project initiation and planning phase (Hoffer, Valacich, and George) of the systems development life cycle for the World Wide Web project at TechPrint. The major deliverable from this phase is the Baseline Project Plan (BPP) and the Statement of Work. The BPP is used by Chad to decide whether the project is to be selected, redirected, or canceled. The Statement of Work should describe what the project will deliver and outline all work necessary to complete the project.

The format of Baseline Project Plan report is shown in Figure 9-1.

BASELINE PROJECT PLAN REPORT

1.0   Introduction

    A.   Project Overview—Provides an executive summary that specifies the project's scope, feasibility, justification, resource requirements, and schedules. Additionally, a brief statement of the problem, the environment in which the system is to be implemented, and constraints that affect the project are provided.

    B.   Recommendation—Provides a summary of important findings from the planning process and recommendations for subsequent activities.

2.0   System Description

    A.   Alternatives—Provides a brief presentation of alternative system configurations.

    B.   System Description—Provides a description of the selected configuration and a narrative of input information, tasks performed, and resultant information.

3.0   Feasibility Assessment

    A.   Economic Analysis—Provides an economic justification for the system using cost–benefit analysis.

    B.   Technical Analysis—Provides a discussion of relevant technical risk factors and an overall risk rating of the project.

    C.   Operational Analysis—Provides an analysis of how the proposed system solves business problems or takes advantage of business opportunities in addition to an assessment of how current day-to-day activities will be changed by the system.

    D.   Legal and Contractual Analysis—Provides a description of any legal or contractual risks related to the project (e.g., copyright or nondisclosure issues, data capture or transferring, and so on).

    E.   Political Analysis—Provides a description of how key stakeholders within the organization view the proposed system.

    F.   Schedules, Timeline, and Resource Analysis—Provides a description of potential timeframe and completion date scenarios using various resource allocation schemes.

4.0   Management Issues

    A.   Team Configuration and Management—Provides a description of the team member roles and reporting relationships.

    B.   Communication Plan—Provides a description of the communication procedures to be followed by management, team members, and the customer.

    C.   Project Standards and Procedures—Provides a description of how deliverables will be evaluated and accepted by the customer.

    D.   Other Project-Specific Topics—Provides a description of any other relevant issues related to the project uncovered during planning.

Figure 9-1   Baseline Project Plan report (Figure 7-10 in Hoffer, et al., *Modern Systems Analysis and Design*, p. 248. © 1996 by The Benjamin/Cummings Publishing Company, Inc.)

Red Rock City

R ed Rock City is a city of 5,000 in the Southwest. The city staff is comprised of 48 full-time-equivalent employees who provide a variety of municipal services for the residents. City Hall houses the following operating departments: city manager, city clerk, finance and planning, and community development. The three other city facilities, including the fire department, community services department, and public works yard, are housed separately from City Hall.

The Red Rock City planning and community development department is interested in improving its management of land-use data. The city council believes that the land-use management information currently available does not allow it to make decisions in the best interest of the city. Since the city incorporated in 1957, it has generated a tremendous amount of data relating to land use. The city council would like to develop a new system that will allow it to manage the basic information about each land parcel and permit application. Additionally, the system should be able to provide references to the archived boxes of permit application files.

LAND-USE BUSINESS ACTIVITIES

A preliminary analysis of the current land-use business activities was undertaken by Julie White, a student intern from one of the local universities. Julie documented the basic permit application process, identified problems with the current operations, and developed a list of data fields needed in a new system, as well as possible outputs from the new system.

The city council, as well as other city employees, use data relating to land usage to make policy decisions. Currently data is not accessible in a timely manner, and reports often contain inaccurate information. This results in decisions that may not be in the best interests of the city. Council members would like to be able to track all information about a land parcel. This information would include the current owner, city zoning codes, pending permit applications, approved permits and denied permits, permit applicants, architects, and staff planners for each application.

The city council also believes that it may be able to generate revenue for the city by having better access to this data. Realtors often request permit histories for a property. The current system prevents easy access to this data. However, an automated system will be able to provide this information quickly and would allow the city to charge a small fee for this service.

As Julie studied the need for a new system, she identified the following problems with the current system:

- Tracking and managing permit applications is difficult. It is hard to determine whether a permit application is simply a resubmission of an already denied application. It is also difficult to monitor the progress of pending applications.

- Cross-referencing of previous permit applications with pending applications is burdensome and often results in inaccurate information.

- Generating required and ad hoc reports is very time-consuming.

PERMIT APPLICATION PROCESSES

When an applicant, usually the homeowner or a general contractor, submits a request for a permit, the public counter staff logs it into one of two log books. One log book is designated for permits that are reviewed by the planning commission, and the other log book is designated for permits reviewed by the design review board. Red Rock City is very concerned with preserving the views of property owners and the distinctive architectural style of their community. Any permits that might result in view infringement or affect the external architectural style must be reviewed by the design review board. The planning commission handles other permits, such as internal design modifications.

Both log books list the permit applications by application date. There is no easy cross-referencing capability, such as determining all applications for a specific parcel, through the manual logging system. Therefore, all requests for information are produced through a method that is labor-intensive and tends to be error-prone. Figure 10-1 shows a sheet from the Planning Commission Log Book. Both books are identical except for the type of permit logged into them. The fields included in the log are the date the application was filed, the case number (also called permit application number), the name of the applicant (applicants could be the owner, an architect, or an agent), the name of the project (often the same as the

			Red Rock City Log Book			
Date Filed	Case Number	Name of Applicant	Name of Project	Location of Property	Related Cases
11-24-91	V-91-11	Catellus Dvlpmnt	Red Rock Train St.	Main & Division	DRB-91-49
1-28-92	V-92-1	Ed Eginton	Burns Residence	3002 Sandy Lane	299-020-05 V-90-23
2-10-92	V-92-02	Cohn & Assoc	Moss Addition	243 Maple	FOP-72-02 299-095-09

Figure 10-1   Planning commission log book

property owner), the location of the property (or street address if available), and other related cases (if known). The related case information may include the parcel number if it is known, or an additional permit number for related work. Sometimes an application may apply to more than one parcel of land. The log book doesn't have a standard way of recording this information. While detailed information is included on the actual application, the office staff sometimes records the additional parcel numbers in the related cases column.

The city also maintains plat maps of property that falls within the city limits. An example of a plat map is shown in Figure 10-2. Red Rock City has an old plat map book that is used at the public counter as a cross-reference tool. Staff planners write the approved permit application numbers within the space of the lot on the parcel map. This process creates several problems. One problem is that this final step, after the permit has been approved, is often skipped. Another problem is the small space available to write in, complicated by illegible handwriting. Additionally, the book is not duplicated anywhere and is very old. The loss of the book would result in an increase in worker time to determine what permits have been approved for a particular parcel.

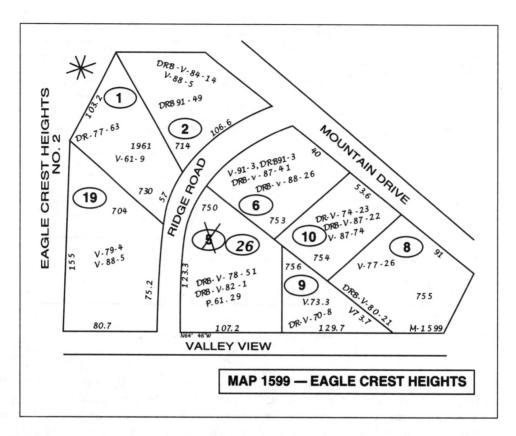

Figure 10-2   Red Rock City plat map

Other data used by the planning and community development department include a listing of all parcels within the city limits (by assessor's parcel number) and the owners' names and addresses. The list is purchased from the county assessor's office once per year. An example is shown in Figure 10-3. The assessor also sends similar listings, one by street address and the other by owners. Additionally, the assessor sends one listing of property by zone ID.

Zones indicate the type of development that can occur on the property. Examples of different zones include R1, R3, and so forth. Each zone has a unique description associated with it, such as low-density residential. Because of the variable size of parcels, it is possible that one parcel may fall under several different zones. Because all city-owned parcels are public property, these parcels are not listed on the assessor's roll. Anytime a listing of public land is requested, a staff member must devote time to retrieving that information by reviewing other city files. These files contain parcel numbers and zoning regulations for the city-owned parcels.

Secured Assessment Roll of Property in the County of Red Rock, Arizona

Owner's Name and Address	Tax Rate Area	Parcel Number	Assessed Values			Exemptions Amount	Net
			Land	Improvements	Personal Property		
Winteroth, Durl 13777 Jacaman, Tucson, AZ 85705	11011	299-242-13-47	5900	600			6500
Williams, Phyllis 3421 Highview Dr., Phoenix, AZ 85257	11011	299-242-13-48	5900	1200			7100
Donnely, Daryl 13709 Mango Dr., Tempe, AZ 85260	11011	299-242-13-49	7548	612			8160
Mareks, Pauline 3350 Vista Ave., Gila Bend, AZ 86301	11011	299-242-13-50	4300	600			4900

Figure 10-3   County assessment rolls

NEW SYSTEM REQUIREMENTS

The council members believe that an automated system will allow them to better manage the permit process as well as assist them in gathering and reporting data relating to land use. They have identified several requirements for their new system. As a minimum, the new system should provide the following capabilities:

- Automated logging of applications. The city must keep the hard copy of the application, and the location should be tracked by the automated system. The location may be with a staff planner, review board, or filed. If filed, the box number should be indicated.

- Automatic reporting of pending applications. Staff must be able to generate reports indicating all pending applications.

- Simplified ad hoc reporting.

- Automatic and accurate parcel history listings. Staff must be able to generate a listing of all approved permits for a parcel.

- Automated listing of business licenses for architects working in the city. All architects who perform work in Red Rock City are required to have a city business license. The city council would like to be able to automatically check that architects listed on the application have a valid license.

The city council would like the system to begin with an automated entry system for the application permit. They would like to modify the current log book information to also include a short description of the project purpose, the staff planner the case has been assigned to, whether the application is for the design board or the planning commission, and a status field. The status could be either pending, approved, or denied. Other information from the application should include the applicant's name, address, and phone number. Similar information should be recorded about the project architect or contractor (one per application). Finally, the council would like to be able to store the assessor's parcel number(s) associated with the permit. The parcel number is a unique field assigned by the county assessor's office.

Julie was able to produce a partial data dictionary of some of the fields (see Figure 10-4), as well as an overview of some basic reporting needs from the new system.

Field	Length/Type	Description
Parcel Number	13, including dashes/ alphanumeric	Unique field to identify parcel
Parcel Address	30/alphanumeric	Street address for a parcel
Permit Number	11, including dashes/ alphanumeric	First 3 characters designate permit type (design review board or planning commission). Next 2 characters designate current year; last four are permit number, e.g., DRB-95-01 or PRC-95-234.
Permit Work Description	30/alphanumeric	Brief description of work to be done on the parcel, e.g., 2,500 sq. ft. addition
Permit Date	8/date field	Date of application for permit
Application Location	20/alphanumeric	Location of the permit application. This can be either the staff planner, the review board or commission, or the filing box number.
Status	1/alphanumeric	Permit status (P, A, D)
Status Date	8/date field	Date of status change from pending
Last Name	25/alphanumeric	Last name for owner, architect, or agent
First Name	20/alphanumeric	First name for owner, architect, or agent
Mailing Address	87/alphanumeric	Mailing address for a parcel owner. Should be broken down into street address, city, state, zip, phone.
License Number	10/alphanumeric	Business license number for an architect. This number is assigned by the city.

Figure 10-4   Partial data dictionary

The reports should be designed to answer the following questions:

1. Listing of properties within a zone showing the Parcel ID, Parcel Address, and Owner ID fields. This would be useful if the city was considering a policy change that would affect an entire zone.

2. Alphabetical listing of property owners showing the Parcel ID, Parcel Address, Owner ID, and Owner Name. This would help in identifying owners of multiple properties.

3. Listing of the status of all permit applications for an applicant. The report should include the following fields: Applicant ID, Applicant Name, Phone Number, Permit #, Description of Permit, Date of Application, Staff Planner Code, Status, and Status Date.

4. Listing of the history of permit applications on a particular parcel. This should include the Permit #, Staff Planner Code, Architect ID Code, and Date of Application.

5. Listing of the applications associated with a particular architect. This would be useful to see the types of projects an architect has worked on or is currently working on. The report should include basic identifying information about the architect (Architect ID, Architect First and Last Name, Phone Number) as well as basic permit information (Permit #, Description of Permit, Date of Application, Applicant ID Code, Status, and Status Date).

6. Listing of the permit applications assigned to a city staff planner. This will be useful for tracking down the most informed staff member about a project. It can also be used by the staff supervisor to help in workload distribution. It should include the basic permit information described earlier.

7. Listing of permit applications for a specified period of time. This will be useful for management to review trends in permit processing and can be used to estimate future development and resulting revenue. It should include the basic permit information described earlier. It may be useful to be able to look at this data based on different fields, such as the status type, permit type, and so forth.

ADDITIONAL INFORMATION

The planning and community development department has recently acquired five personal computers. These computers are 486DX-66 machines with 8MB RAM, 400MB hard drives, and SGVA monitors. They came preloaded with Microsoft Office Professional. Two of the machines will be located at the staff planners' desks, two will be located at the public counter for use by the public counter staff, and the last one will be located at the public counter for use by the public to access public information, such as a listing of all approved permits for a particular parcel, or to identify the owner of a parcel. The staff is not very knowledgable about computers and will need to be provided with basic computer literacy skills as well as training for the new system.

Exercises

1. How would you recommend that the planning and community development department address their concerns about the plat map log book?

2. Develop a detailed system design that will adequately meet the requirements set by the city council. You should develop an entity-relationship diagram, database file design, screen layouts, report layouts, and an implementation plan.

3. Prepare a summary report to the city council outlining your proposed system design and implementation plan.

4. The city has approved your proposal. Develop a prototype of the new system.

5. The county assessor's office has recently notified all incorporated cities that they will be able to obtain the parcel listings in a comma-separated values format. Design an interface module that will permit easy loading of this data into your system.